Water Resources Development and Management

Indexed by Scopus

Each book of this multidisciplinary series covers a critical or emerging water issue. Authors and contributors are leading experts of international repute. The readers of the series will be professionals from different disciplines and development sectors from different parts of the world. They will include civil engineers, economists, geographers, geoscientists, sociologists, lawyers, environmental scientists and biologists. The books will be of direct interest to universities, research institutions, private and public sector institutions, international organisations and NGOs. In addition, all the books will be standard reference books for the water and the associated resource sectors.

Yangju Tu

Management of Hydropower Enterprises

Intelligent Operation, Exploration
and Practice in China's Dadu River Watershed

 Springer

Yangju Tu
CHN Energy Group
Dadu River Hydropower Development Co.,
Ltd.
Chengdu, China

ISSN 1614-810X   ISSN 2198-316X (electronic)
Water Resources Development and Management
ISBN 978-981-97-5583-7   ISBN 978-981-97-5584-4 (eBook)
https://doi.org/10.1007/978-981-97-5584-4

This work was supported by Dadu River Hydropower Dev. Comp..

This Springer imprint is published by the registered company Springer Nature Singapore Pte Ltd.
The registered company address is: 152 Beach Road, #21-01/04 Gateway East, Singapore 189721, Singapore

Foreword

In October 2018, we were invited to the 60th anniversary celebration of the China Institute of Water Resources and Hydropower Research which is the world's largest water research and development institution and belongs to the Ministry of Water Resources.

Like all meetings, this had a few truly outstanding lectures with new and innovative ideas. From our perspective, the most exciting presentation was by Mr Tu Yangju, who was one of the senior-most officers of the Dadu River Hydropower Authority. In a short presentation of 30 minutes, Mr Tu discussed how the Dadu River Hydropower Authority was very effectively using sensors, robotics, big data analytics, advanced models, and artificial intelligence to manage a cascade of dams on the Dadu River, one of the tributaries of the Chang Jiang (Yangtze River). It became clear to us this state-owned enterprise was at least 5 years ahead of any of its counterparts, either in China or in the rest of the world.

During the break in the session, we went directly to Mr Tu, introduced ourselves, and wanted to know more about what the Dadu River Hydropower Authority has achieved and what are its future plans in terms of research and development activities.

Mr Tu was gracious enough to spend some time with us explaining what they had done and their plans for the future. Coincidentally, our Chinese host, the Institute of Water Resources and Hydropower Research, had planned for us to fly to Chengdu for some additional activities immediately after their anniversary celebrations. Most fortunately, and coincidentally, Chengdu is the headquarter of the Dadu River Hydropower Authority. When Mr Tu became aware of our plan, he cordially invited us to spend a couple of extra days in Chengdu so that we could visit one of their dams and see for ourselves how they are using advanced and innovative technologies to significantly increase the outputs of their hydropower plants.

After visiting one of their major hydropower plants, we were truly impressed by the tremendous advances that the Dadu River Hydropower Authority had made in using new techniques to significantly improve the operating efficiencies of the plant. Mr Tu gave us a copy of his book *Intelligent Operation and Management of Hydropower Enterprises*. The book was in Chinese, and, most unfortunately, our Chinese reading proficiency leaves much to be desired!

On our return to Singapore, we gave the book to one of our Chinese colleagues and requested him to read it and give us his objective views on it. In about 3 months, our colleague said it was a truly excellent book, and he briefly described what were the issues he found most exciting and new. After listening to him, it was clear to us that the book should be made available to the English-speaking world so that water experts learn from these remarkable advances. We thus approached Mr Tu to see if he would be interested in having the book translated into English. We also offered to him that if he could have it translated, we would get one of the world's most reputable technical publishers, Springer Nature, to have it published promptly in English. Mr Tu enthusiastically responded positively to this suggestion. The current book is not only a translation but also has been updated with new materials.

Since our first meeting in 2018, in Beijing, we have met with Mr Tu in many cities in China, including Chengdu. After each meeting, we have returned with not only steadily increasing respect for Mr Tu and the excellent work of his staff but also the steady advances Dadu River Hydropower Authority has been making. Our latest meeting was in September 2024, in Beijing, again through the courtesy of the China Institute of Water Resources and Hydropower Research.

We are confident that hydropower professionals from all over the world will get many new ideas as to how they can further improve the operating efficiencies and management practices of their plants.

Asit K. Biswas
Academician and Distinguished
Visiting Professor
University of Glasgow
Glasgow, UK

Cecilia Tortajada
Professor in Practice School of Social
and Environmental Sustainability
University of Glasgow
Glasgow, UK

Preface

This book was produced over a period of four years and received the encouragement and guidance of Prof. Asit K. Biswas[1] and his wife, Dr. Cecilia Tortajada.[2] Here, I would like to express my great respect to them both!

Looking back, the book preparation process was a wonderful experience, especially the interactions with Prof. Biswas and his wife, Dr. Tortajada, and I would like to share a bit of this story with the readers.

---

[1] Professor Asit K. Biswas, Canada, is primary engaged in water resources and environmental management research, and is an internally recognized authority in the field of water environment management. In the 1980s, he was invited by Chinese leader Deng Xiaoping to China to consult on the South-North Water Diversion Plan. A Distinguished Visiting Professor at the Lee Kuan Yew School of Public Policy at the National University of Singapore, and a Distinguished Visiting Professor and Senior Advisor to the Vice-Chancellor at the University of Glasgow, UK, he is one of the founders of the International Water Resources Association and the World Water Council, a former President of the International Water Resources Association, senior advisor to 20 countries and Governments, and to six heads of United Nations agencies, and Secretary-General of several international organizations and agencies. Professor Biswas won the Stockholm Water Prize, which is known as the Nobel Prize in the field of water research, as well as government awards from Canada, Spain, the United Kingdom, and India, and was named one of the "Top 10 water trailblazers" by Thomson Reuters.

[2] Dr. Cecilia Tortajada, Mexico, Senior Adjunct Research Fellow at the Institute of Water Policy at the Lee Kuan Yew School of Public Policy, National University of Singapore, and professor at the University of Glasgow, UK. An internationally renowned expert in water resources, environment, and agricultural management, she is mainly engaged in inter-disciplinary research in water, environment, food, and energy securities. She currently serves as an advisor to several United Nations agencies, such as the United Nations Environment Programme, the Food and Agriculture Organization of the United Nations, the Canadian International Development Research Centre, the World Bank, the Asian Development Bank, and other international organizations, as well as the governments of more than 10 countries, including Spain, Germany, and Brazil. She serves as the editor-in-chief of the International Journal of Water Resources Development and is the associate editor and editorial board member for several international SCI journals.

Fig. 1 The author meeting with Prof. Biswas and Dr. Tortajada in the city of Chengdu, Sichuan Province, China

First Meeting

On October 17, 2018, I was invited to participate in the 60th anniversary celebration of the establishment of the China Institute of Water Resources and Hydropower Research (CIWRHR), and I gave a presentation regarding the intelligent hydropower enterprise at the International Symposium on Water System Operations. Prof. Biswas and Dr. Tortajada also attended this symposium, where Prof. Biswas gave a keynote presentation as well. The couple were quite interested in my report and were able to get my contact information through the conference organizers. Several days later, on October 21, Prof. Biswas was in Chengdu to discuss a special issue of the *International Journal of Water Resources Development*, and we had our first meeting. During our encounter, he and I discussed artificial intelligence application scenarios for hydropower enterprises, and he showed great interest in the application of new technologies such as big-data and artificial intelligence in hydropower enterprises, and was impressed by the achievements of the Dadu River Company in intelligent enterprise development. We talked with great pleasure. As we were about to part, I gave him a copy of my book *Intelligent Enterprise—Framework and Practice,* published in 2016. Naturally, however, I was unsure whether Prof. Biswas, who did not know Chinese, would be willing to spend the time to become familiar with such a book in Chinese.

Initial Beginnings

With that parting I was not sure when I might again be in contact with Prof. Biswas and Dr. Tortajada. Then, three months later, I unexpectedly received an email from the professor in Singapore. I was pleasantly surprised by the news in the professor's email. It turns out that the professor had given my book to one of his Chinese colleagues to read, and after reading the book, with excitement the colleague immediately described the content of the book to the professor, telling him that it was a very good book, and that he had never seen a similar book published in English. The professor emailed me and suggested that I could start writing a book in English on the application of artificial intelligence technology in hydropower management. He thought that such a book would be an excellent addition to the knowledge of intelligent hydropower operation and management for the English-speaking world, and would certainly be very well received in the English-speaking world.

The high level of acknowledgement from Prof. Biswas was truly exciting. He believed that China was far ahead of the rest of the world in this emerging field. It would be a very exciting thing if a book could be published in the English-speaking world on the latest developments in Chinese hydropower, but it was also easy to imagine how difficult it might be.

Just as I was hesitating, Prof. Biswas sent a second email, just two days after the first. The professor again emphasized that hydropower professionals in the English-speaking world would greatly appreciate a book on such a topic, especially since China was far ahead of the West in the use of big-data and artificial intelligence in hydropower management. The professor's reply certainly gave me a boost of confidence, and my sense of mission in this matter would be sufficient to help me overcome any hardship. So I made the decision to write and initially drafted two possible book titles for comment. Prof. Biswas replied immediately with an email suggesting the title *Intelligent Operation and Management of Hydropower Enterprises.*

Unexpected Interruption

As I was beginning to write the book, at the end of February, I received notice that my superiors had arranged for me to go to Beijing to participate in training for nearly half a year. The writing work would have to stop. Towards June, I had no choice but to convey to Prof. Biswas the news of the delay of the manuscript.

After becoming familiar with the situation, Prof. Biswas expressed his understanding at the manuscript delay and methodically made subsequent adjustments. The professor also expressed his willingness to proofread the manuscript together with his wife, Dr. Tortajada, prior to publication.

In that same email, the professor started talking to me about matters other than the business of writing the book. He told me that he and his wife would be coming back to China in early September to participate in the Global Sustainable Development

Forum, in Kunming, organized by the Yunnan Provincial Government. He and Dr. Tortajada would each be giving keynote speeches, and that they were both very much looking forward to meeting with me. The professor also talked to me about his affection for China. In 1980, Chinese leader Deng Xiaoping had invited Prof. Biswas to China to evaluate the South-North Water Diversion Plan. This was the professor's first visit to China, and he fell in love with China during his six-week visit. In the subsequent 40 years, he visited China at least once a year, sometimes even three times a year.

I was honored to have been able to meet such an international friend, and touched by the professor and his wife's deep affection for China. After that, we continued corresponding with each other at least once a month, speaking in our emails about more things other than writing books, and gradually building a deep friendship.

Lakeside Reunion

At the end of August 2019, I received an email from Prof. Biswas that he and his wife would be visiting Kunming during September 2–6, 2019. Almost a year had passed since my first meeting with the professor, and I was very much looking forward to seeing him again so as to learn about recent trends in global sustainable development, to discuss together the latest ideas on intelligent hydropower management, and to confer on a subsequent work plan. So I immediately arranged my itinerary and headed towards to Kunming to meet with the professor.

On September 5, 2019, I again met Prof. Biswas and his wife, Dr. Tortajada, this time on the shores of Fuxian Lake, about 70 minutes away from the Kunming Airport. I introduced some of our new achievements from the last year related to intelligent hydropower operation and management, and the professor could not wait to see these results soon presented in the book. We then had an in-depth discussion on manuscript writing plans, along with other matters. I introduced the outline of the book, the challenges encountered in the process of writing, and the next step of the writing plan. The professor and Dr. Tortajada both gave me much encouragement and good advice, and this meeting made me feel more confident about the upcoming writing work.

Meeting Again In Nanjing

Just one month after the lakeside reunion, Prof. Biswas received the good news that he would come to China again, and he would be in Nanjing from December 16 to 24, at the invitation of the scholar Tang Hongwu, of Hehai University. The professor hoped he could meet with me again at that time.

On December 18, while attending a national intelligent enterprise summit in Nanjing, I took advantage of the opportunity to meet with Prof. Biswas as arranged.

Fig. 2 The author meeting with Prof. Biswas and Dr. Tortajada in the city of Kunming, Yunnan Province

This time, we had an in-depth discussion on the book's English language abstract and outline, with a focus around the position of people in the intelligent enterprise in an age of artificial intelligence. Prof. Biswas said that China was ahead of the rest of the world in sensors, artificial intelligence, robotics, and big-data analysis, and that it was remarkable that the Dadu River Company had trained some of its personnel for reorientation to employment in the technology sector, thus preempting the social problem of unemployment caused by the application of new technologies such as artificial intelligence.

Conference Regrets

After the meeting in Nanjing, on January 4, 2020, I received an invitation email from Prof. Biswas to attend the Kuala Lumpur International Hydropower Conference, which would be held in Kuala Lumpur, Malaysia, on March 10–12, 2020. Around 200 hydropower experts from different parts of the world, as well as experts from the World Bank and the Asian Development Bank, would attend such an unparalleled conference. The UK organizers had asked Prof. Biswas to deliver the opening address, on March 10, on the future of hydropower in the world after 2020. The

Fig. 3 Scholar Tang Hongwu of the Chinese Academy of Engineering, and the author, meeting with Prof. Biswas in Nanjing

professor proposed to the organizers to add a lecture by me to the opening session, for me to share about the application of technologies such as artificial intelligence and data analytics in hydropower management, and he indicated that the organizers had expressed their strong support for this. Even if I could not attend the conference in person, Prof. Biswas wanted me to convert the main content of the book into a talk and send it to him, and he would make sure to share it at the conference.

Although it was a last minute invitation, it would be a great personal honor for me to present the latest achievements of Chinese hydropower enterprises to the world at such an event, and I replied in the affirmative. The professor was also very excited about my participation in the conference. Shortly later, I received a formal invitation from the UK organizers. I was also excited to hear that the professor and Dr. Tortajada would come to Chengdu for a week in the beginning of April.

Then came the end of January 2020, a point in time at which the whole world shuddered, the sudden outbreak of the novel coronavirus epidemic shattering the Chinese New Year Eve (January 24, 2020). Throughout the February following, China was shrouded within the gloom of the epidemic, and I began to worry that the International Hydropower Conference, and my engagement with Prof. Biswas, would be affected. Then, on the last day of February, I received an email from Prof. Biswas, and my worry was confirmed. Due to the epidemic, the Kuala Lumpur conference would be postponed, and any rescheduling was uncertain.

Passing Through the Pandemic Together

The novel coronavirus epidemic was a global catastrophe, with more than 100,000 confirmed cases worldwide by March 2020. So the epidemic went from a one-way concern of Prof. Biswas for my well-being, to a topic which we discussed together. In the year that followed, we had many exchanges regarding status of epidemic control in China and elsewhere. Prof. Biswas and Dr. Tortajada also carried out research on epidemic control in China, giving their analyses and writing articles from their professional points of view, and I was fortunate to be one of his first Chinese readers of these articles.

On March 19, 2020, the professor sent me an article on novel coronavirus that he and his wife had written at the invitation of *China Daily*, in which he analyzed the advantages and results of China's policies to control the virus, and also attempted to compare these policies to the attempts to control the SARS outbreak in 2003, expressing his own opinions. I was impressed by the degree of the professor and his wife's knowledge of China.

On June 25, 2020, the professor and his wife sent two more articles. One had been written at the invitation of the China Daily, with views on how Beijing was dealing with the second coronavirus wave. The other was an article on how the coronavirus outbreak might affect water supply, wastewater management, and sustainability goals. This article was published globally in eight languages (including Chinese) by Project Syndicate. Reading it, I once again sensed the professor and his wife's high level of professionalism and their broad vision.

The Dust Settles

The first draft of the book was completed in July 2020, but I was always dissatisfied with it, so I revised it again and again, until finalizing the manuscript a year later, and in August 2021, with the dust finally settled on the Chinese version, I passed that news right away to Prof. Biswas, who was very excited.

In view of global epidemic circumstances, we came up with a "two-legged" approach" of Chinese and English versions in parallel, the Chinese version to be released in China first, and then after translation, an English version to be released to the English-speaking world. Time has flown by, almost a year has passed, and with the considerate help of Prof. Biswas and his wife, the English version is finally available to you all.

Correspondence Friendship

Originally, I should have finished writing at this point. But while going through the back and forth email correspondences, the friendship embedded in the exchanges with Prof. Biswas jumped out at me. Just some small exchanges, really, but all about our shared friendship. I could not bear to stop writing, so I organized some of the original texts to share with readers.

July 16, 2020: "We hope the current floods in China are manageable in the Chengdu area, and also in the Dadu River Basin. We notice already water levels in many areas have broken historic records, and higher even than the 1998 catastrophic floods. At the request of China Daily Cecilia and I wrote our views on the current flood situation. This article is attached and may be of some interest to you."

October 15, 2020: "There have suddenly been some new developments in our life. Cecilia has accepted a Research Professorship at the University of Glasgow. I already have a position as Distinguished Visiting Professor and Senior Advisor to the Vice Chancellor at the same university. While I can carry out my duties remotely with two annual visits to Glasgow, Cecilia's appointment means we have to move to Glasgow. Thus, we are planning to move to Glasgow during the middle of November. This will be a new adventure. I started my professional career in Glasgow over fifty years ago. In that sense, it would be great to go back again."

October 31, 2020: "Cecilia has accepted an offer from the University of Glasgow… to be a full research professor at its School of Interdisciplinary Studies. I shall be an advisor to the Vice Chancellor of the University. Thus, on 20 November, we are leaving Singapore, and moving to Glasgow."

April 13, 2021: "Last week, China Daily invited Cecilia and I to write a piece outlining our views as to how China eradicated absolute poverty. They published it yesterday. It is attached for your possible perusal. We are pleased that it was the most viewed piece on Monday."

July 10, 2021: "We have not corresponded for some time. We trust everything is going well with you, your family and all the members of your staff. Here in the UK, the Government has not handled the COVID-19 infections as well as in China. Fortunately, both Cecilia and I are now fully vaccinated… At the request of China Daily, we gave our views on the possibility of China becoming carbon neutral by 2060. This piece is attached."

Chengdu, China   Yangju Tu

Contents

Chapter 1
The Dadu River Watershed

1.1 Rich Hydropower Resource Potential

The Dadu River is the largest tributary to the Minjiang River, itself a tributary river to the upper Changjiang River, a vast and important river that stretches all the way to China's eastern seaboard. The Dadu River originates in the south foothills of the Guoluo mountain on the high plateau of Qinghai Province and has two sources—the western source being the Chuosijia River and the eastern source being the Zumuzu River, this latter being the main flow. The complete mainstem of the Dadu River is 1062 km long, flowing roughly from north to south initially, passing through counties in Sichuan such as Jinchuan County, Danba County, and Luding County, then at Shimian County turning to the east, and flowing through places such as Hanyuan County, Ebian County, and Shawan District, before absorbing the Qingyi River, and then finally pouring into the Minjiang River at Leshan. The total area of the watershed is 77,700 km^2.

The Dadu River watershed (shown in Fig. 1.1) is located in the transition zone from the southeast edge of the Qinghai-Tibet Plateau to the western part of the Sichuan Basin. The watershed is bordered on the north by the Bayankala mountains and the Yellow River. On the south, it is adjacent to mountains like the Xiaoxiang Range, Daliang Mountain, as well as the Jinsha River. To the east, it is bordered by Zhegu Mountain, Jiajin Mountain, the Daxiang Range, and the Minjiang River and the Qingyi River. To the west, it is neighbor to Luokema Mountain, Dangling Mountain, Zheduo Mountain, and the Yalong River. The watershed is surrounded on every side by lofty mountains, the perimeter elevation generally above 3000 m, with many peaks of 4000–5000 m. The boundary mountain range in the middle reaches of the river has some passes with relatively low elevations, of around 2000 m, and these act as a main passage for atmospheric water vapor.

The Dadu River watershed spans five latitudes and four longitudes. Additionally, a large range of altitudes, along with complex terrain, result on great variation in climate across the watershed. The upper river has subarctic and cold temperate climates,

Y. Tu, *Management of Hydropower Enterprises*, Water Resources Development and Management, https://doi.org/10.1007/978-981-97-5584-4_1

Fig. 1.1 Geographic elevation map for the Dadu River

with distinct wet and dry seasons, a long winter and no summer, and yearly average temperature of 6–12 °C. The high altitude and long-distance from sources of water vapor result in relatively limited precipitation, with average amounts many years being only around 700 mm. The middle reaches of the river have a subtropical humid climate, and although changes in altitude still result in noticeable climate variation, the river valley has four obvious seasons. The yearly average temperature is 13–18 °C, and most years the average annual precipitation is 700–1200 mm. The lower river is characterized by warm winter, hot summer, cool autumn, and a relatively humid climate, with an ample supply of water vapor, and abundant precipitation. Many years, the average precipitation is 1300–2500 mm. According to statistics based on hydrological station observational data, the typical average annual flow of the Dadu River watershed is 1500 cubic meters per second, and the typical annual average runoff is 47.3 billion cubic meters, comparable to the China's Yellow River.

The combination of tremendous gradient, abundant water volume, and narrow river valley together form abundant resources for hydropower development in the Dadu River watershed. The theoretical hydropower reserves in the whole watershed, starting at the origin in Qinghai Province and extending through the greater watershed in Sichuan Province, are as high as 31.32 million kW, ranking fifth among the thirteen major hydropower bases in China, with the hydropower potential just within Sichuan Province itself reaching 30.12 million kW, accounting for 20.6% of the total cumulative hydropower resources of the various rivers in Sichuan Province. Most remarkable is the 593 km segment of the mainstem from Shuangjiangkou to Tongjiezi, with a natural drop of 1827 m elevation, and hydropower reserves of 17.48 million kW, representing more than 50% of the potential capacity of the entire watershed. In addition to the abundant hydropower resources of the mainstem, the six tributaries, Zhuosijia River, Xiaojin River, Wasigou, Tianwan'gou, Nanya River, and Niri River, all have potential capacities exceeding 500,000 kW.

1.2   A Long Human History

Beautiful verdant mountains and lovely scenery follow the line of the Dadu River, and the warmth of its winters and unique physical geography create conditions suitable for the existence of diverse species and human settlements. Forests, grasslands, and minerals are all relatively abundant, with the area of the original native forest on western side on the watershed making up 15.3% of Sichuan Province's total forested area, with 26.1% of Sichuan's lumber stocks. This forest is abundant with every kind of fruit, medicinal plant, and secondary forest product, ranging from subtropical to temperate, and these enjoy an excellent reputation on the international market. The rich natural resources nurture the sons and daughters of the multi-ethnic Dadu River, and various groups such as Han, Tibetan, Yi, and Qiang, dwell along the path of the watershed, which contains China's second largest Tibetan settlement area, the largest concentration of Yi settlements, and the only Qiang ethnic settlement area, forming a blended cultural corridor of multiple ethnic groups, as shown in Fig. 1.2.

Fig. 1.2 A vignette of natural scenery on the Dadu River

The many well-known cultural attractions and the numerous natural landscapes complement each other well, with famous Buddhist holy places such as Mount Emei (Fig. 1.3), and the Leshan Giant Buddha, the world's largest carved-stone seated Maitreya Buddha (Fig. 1.4), as well as the "king among mountains," Mount Gongga.

The famous Buddhist mountain, Mount 4mei, located on the banks of the lower reaches of the Dadu River, has a maximum altitude of 3099 m and is located in a region where many natural factors overlap and interact. The mountain has as many

Fig. 1.3 On Jinding, Mount Emei's highest peak

Fig. 1.4 At the meeting place of Dadu River and Minjiang River: The Leshan Giant Buddha

as 3200 species of plant and more than 2300 species of animal, variously known for being fierce, beautiful, fantastical, and dangerous, and uniquely wonderful. There are many historical sites and temples on Mount Emei, and the mountain is rich in cultural relics, many of which are rare treasures. All the buildings, statues, ceremonial musical instruments, and imagery reveal a rich Buddhist religious atmosphere.

About 40 km from Mount Emei, on a cliff at Leshan City's Lingyun Mountain, there is a 71-m-tall stone-carved Maitreya Buddha, that is, the world-famous "Leshan Giant Buddha." The scenic area anchored by the Leshan Giant Buddha, and along with the Mount Emei scenic area, together form a World Cultural & Natural Heritage Site. The Leshan Giant Buddha is located at the confluence of the Dadu River, the Minjiang River, and the Qingyi River. According to legend, 1300 years ago the force of the water during flood periods was quite savage at this place, frequently causing the tragic loss of life and the destruction of boats. The Buddhist disciple Hai Tong held a faith that the floods could be calmed and the people saved by the power of the Buddha. Through many trials, and utilizing the resources and efforts of many, the work was started. Eventually, through the work of three generations of craftsmen stretching over 90 years, the Giant Buddha was finally built. Today, the Leshan Giant Buddha still stands at the confluence of the three rivers, full of the wonderous flavor of ancient Eastern Buddhism.

The most fantastical sight in the Dadu River watershed in the snow-covered Mount Gongga (Fig. 1.5), the name of the mountain in Tibetan meaning "the highest snow mountain." It is the main peak among the snow-covered peaks of the Hengduan Mountain range, with an altitude of 7556 m, appearing in the distance as a vast pyramid towering above the gathered peaks. For climbers, Mount Gongga has an

Fig. 1.5 The king among mountains: Mount Gongga

unparalleled appeal, and this has earned it the reputation of "king of the mountains." The Mount Gongga scenic area is comprised of locations such as Hailuogou, Yanzigou, Mugecuo, Wushuhai, Gongga South Slope, and is a national-level scenic area. Mount Gonga resides in an ethnic minority area, and in the area are Buddhist temples such as the Gongga Temple, the Tagong Temple, and tourists can get a taste of the colorful ethnic customs of the Tibetan and Yi peoples.

This beautiful and richly endowed river is also a necessary route for China's ancient Southwest Silk Road and the TeaHorse Ancient Road, and a location contended over by the military strategists of each dynasty. It is full of historical and cultural heritage and a feeling of legend. Many much-relished historical stories take place here: Zhuge Liang's "Seven Captures of Meng Huo," Qianlong's "Pacification of Jinchuan," and the tragic and stirring demise of the Taiping Rebellion's leader Shi Dakai. In more recent times, the courageous performance of the Red Army on the Dadu River has written a glorious page in the history of the Chinese revolution: The Red Army's desperate last-ditch battle at Anshunchang, and the heroic seizing of the Luding Bridge (Fig. 1.6), which are memorialized in epic poem as "climbing snowy mountains, and passing through grasslands."

Fig. 1.6 The chain spans of the Luding Bridge

1.3 Beginnings of Hydropower Development

1.3.1 Watershed Planning

In order to develop and utilize the abundant hydropower resources of the Dadu River watershed, in the 1950 the relevant authorities in China started survey, planning, and design work for the watershed, defining the development mission for the mainstem as primarily for hydropower, along with flood prevention, transport, and irrigation. As of the end of 2020, the plan for the mainstem included three reservoirs and 38 cascade power stations, with Xiaerxia reservoir as the "faucet" reservoir at the top, Shuangjiangkou reservoir as a key regulating reservoir for the upper river, and Pubugou reservoir as the key regulating reservoir for the middle river section. The total installed capacity for the watershed is around 27 million kW, with a designed total yearly generation capacity of around 116 billion kWh, representing 24% of Sichuan Province's total hydropower resources, this yearly output comparable to saving 35.26 million tons of standard coal equivalent, reducing yearly CO_2 output by 92.38 million tons. Development of hydropower in the watershed has great significance in regard to power infrastructure for the whole of Sichuan Province, as well as for ecological protection along the course of the watershed, and coordinated economic and social development for the region. Figure 1.7 shows the cascade hydropower development plan for the Dadu River watershed.

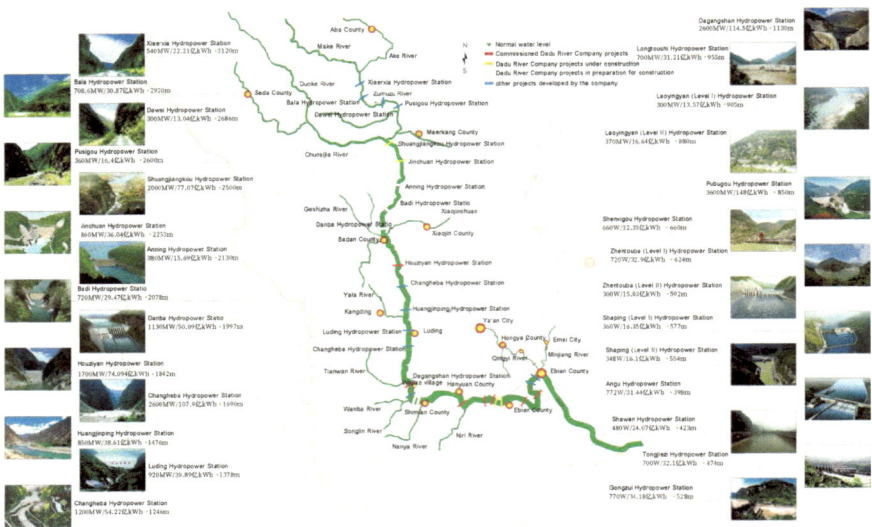

Fig. 1.7 Cascade hydropower development plan for the Dadu River watershed

1.3.2 Development and Commissioning

The superior development conditions of the Dadu River watershed are defined by the river's unique physical geography and large quantity of high-quality hydropower resources. At the same time, given the nearby load center of the Sichuan power grid, with most power stations around 200 km in straight line distance from the city of Chengdu, the mainstem of the river has a natural advantage in terms of supplying power to the Sichuan power grid load center.

In 1966, the first battle of hydropower development in the Dadu River watershed was officially sounded at Leshan City's Gongzui power station. After several years of hard struggle during a time that relied primarily on human labor for construction, by 1971 a concrete gravity dam of 85.5 m height had been built above the surging turbulent waters of the Dadu River, with a total installed total capacity of 700,000 kw, which played an important role in the safe and stable operation of the Sichuan power system. Through the work of multiple generations of hydropower builders on the Dadu watershed over almost half a century, by the end of the 2020, 14 hydropower stations had been built on the mainstem of the Dadu River, with an installed total capacity of around 17.44 million kW, each year delivering on average about 75.7 billion kWh of clean electricity to society. Additionally, construction work had already started on power stations at Shuangjiangkou, Jinchuan, Yingliangbao, Zhentouba (Level II), and Shaping (Level I), and preliminary work on the remaining nine power stations had begun, creating steady development momentum via the

Fig. 1.8 Pubugou hydropower station

progression of project commissioning, projects under construction, and construction preparation for new projects.

A number of landmark hydropower projects have been built on the Dadu River, including dams such as the "International Milestone Project," the 186 m tall Pubugou gravel-soil core rockfill dam (completed and commissioned in 2009; see Fig. 1.8.); The dam built to the highest earthquake resistance standard in the world, the 210 m Dagangshan arch dam (commissioned in 2015; see Fig. 1.9); The world's tallest dam, currently under construction, the 315 m Shuangjiangkou gravel-soil core rockfill dam (planned for commissioning and power generation by 2025). In addition to these landmark projects, there is a wide variety of types of hydropower generator sets on the Dadu River, and among the commissioned sets on the Dadu River watershed, forms include the Pelton turbine, Francis turbine, Kaplan turbine, and bulb-tubular turbine, among others. For this reason, the Dadu River has been called a "hydropower museum" within the industry.

1.3.3 Operations Management

After reform of China's national power system, the original mode of having a single development owner for a whole watershed was disrupted, and multiple developers started developing projects on the Dadu River mainstem, as detailed in Table 1.1. The Dadu River Hydropower Development Co., a subsidiary under the banner of

Fig. 1.9 Dagangshan hydropower station

the China Energy Investment Corporation (China Energy), was primarily responsible the development and operations management for 17 cascade power stations on
the mainstem, with a total installed capacity of about 17.6 million kW, accounting
for about 70% of the planned installed capacity of the main stem. By the end of
2020, the eight power stations on the mainstem under development responsibility of
the Dadu River Hydropower Development Co., Houziyan, Dagangshan, Pubugou,
Shenxigou, Zhentouba (Level I), Shaping (Level II), Gongzui, Tongjiezi, had already
been commissioned, with a total installed capacity of 11.1 million kW. Eleven other
power stations on the mainstem were invested and developed by enterprises such
as China Huaneng Group, China Datang Corporation, Power Construction Corporation of China. Among these power stations, the six commissioned power stations of
Changheba, Huangjinping, Luding, Longtoushi, Shawan, and Angu, have an installed
capacity of 6.342 million kW.

By the end of 2020, in addition to the 17.44 million kW of installed hydropower
capacity in operation on the mainstem of the Dadu River, the installed capacity of
the tributaries of the Dadu River was 4.42 million kW, with the combined capacity
of the mainstem and tributaries reaching 21.86 million kW, representing 36% of the
installed capacity of the Sichuan power grid. At the same time, large-scale power
stations, such as Pubugou, Dagangshan, and Houziyan, played the role of peak and
frequency regulation for the Sichuan power grid, ensuring the smart cooperative
control and safe and stable operation of the power grid. These dams have been
dubbed Sichuan power supply's "stabilizers and ballast stones."

Because of the special status of Dadu river hydropower in the Sichuan power
grid, requirements for its safe, scientific, and economical operation are even higher.
For this reason, the Dadu River Company readily answers every challenge, actively
embracing the development of new technologies, new processes, and new materials,

Table 1.1 Dadu River watershed hydropower station planning and development status

Table order	Project	Location	Distance from dam (gate) to mouth of river (kilometers)	Regulating Capacity	Installed Capacity (10,000 kW)	Construction Status	Developer
1	Xiaerxia	Aba	797	Multi-year	54	Preliminary work	Power Construction Corporation of China: Hydropower Company
2	Bala	Maerkang	766		56	Preliminary work	
3	Dawei	Maerkang	748		36	Preliminary work	
4	Busigou	Maerkang	700		30	Preliminary work	China Energy Investment Corporation: Sichuan Company
5	Shuangjiangkou	Maerkang, Jinchuang	650	Yearly	200	Under construction	China Energy Investment Corporation: Dadu River Hydropower Development Co
6	Jinchuan	Maerkang, Jinchuang	616	Daily	86	Under construction	
7	Anning	Jinchuan	580	Daily	38	Preliminary work	
8	Badi	Danba	545	Daily	72	Preliminary work	
9	Danba	Danba	528	Daily	119.66	Preliminary work	
10	Houziyan	Kangding, Danba	468	Seasonal	170	Completed	
11	Changheba	Kangding	423	Seasonal	260	Completed	China Datang Corporation: Sichuan Company
12	Huangjinping	Kangding	407	Daily	85	Completed	

(continued)

Table 1.1 (continued)

Table order	Project	Location	Distance from dam (gate) to mouth of river (kilometers)	Regulating Capacity	Installed Capacity (10,000 kW)	Construction Status	Developer
13	Luding	Luding	375	Daily	92	Completed	China Huadian Corporation: Sichuan Company
14	Yingliangbao	Luding	351	Daily	111.6	Under construction	China Huaneng Group: Sichuan Company
15	Dagangshan	Shimian	314	Daily	260	Completed	China Energy Investment Corporation: Dadu River Hydropower Development Co
16	Longtoushi	Shimian	294	Daily	72	Completed	Longtoushi Hydropower Company
17	Laoyingyan (Level I)	Shimian	275	Daily	22	Preliminary work	China Energy Investment Corporation: Dadu River Hydropower Development Co
18	Laoyingyan (Level II)	Shimian	263	Daily	35	Preliminary work	
19	Pubugou	Hanyuan	194	Yearly	360	Completed	
20	Shenxigou	Hanyuan	177	Daily	66	Completed	
21	Zhentouba (Level I)	Jinkouhe	152	Daily	72	Completed	
22	Zhentouba (Level II)	Jinkouhe	148	Daily	32.6	Under construction	
23	Shaping (Level I)	Ebian	142	Daily	38	Under construction	
24	Shaping (Level II)	Ebian	129	Daily	34.8	Completed	
25	Gongzui	Leshan	93	Daily	77	Completed	

(continued)

Table 1.1 (continued)

Table order	Project	Location	Distance from dam (gate) to mouth of river (kilometers)	Regulating Capacity	Installed Capacity (10,000 kW)	Construction Status	Developer
26	Tongjiezi	Leshan	65	Daily	70	Completed	
27	Shawan	Leshan	50	None	48	Completed	Power Construction Corporation of China
28	Angu	Leshan	15	Daily	77.2	Completed	

while vigorously promoting digital transition, and widely applying big-data and artificial intelligence (AI) technology, creating an entirely new approach to intelligent operation and management.

Chapter 2
Challenges Facing Operation and Management

2.1 Challenges Facing Production Operation

2.1.1 Complexity of Hydrometeorology

As mentioned previously, the Dadu River watershed is encircled by lofty mountains on all sides, and the terrain is complex. There is great variation in climatic features and physical geographic characteristics across the upper, middle, and lower reaches, with uneven water conditions across different times and places, large variations in flow volume, and poor flood carrying capacity. Floods within the watershed are primarily formed from precipitation, and there are typically 100–170 precipitation days yearly. In the middle and lower reaches, in some areas, precipitation can exceed 180 days yearly. In the upper river, precipitation is characterized by large volumes and extended durations, with a single flood process lasting 5–7 days. Precipitation over a large area or lasting an extended period can result in exceptionally large flooding. In the middle and lower river, heavy rainstorms are common, peaking at over 1000 mm of precipitation in a single day. Since the crisscrossing high mountains are not conducive to the dispersion of rainstorms across space, this results in relatively frequent localized rainstorms in some areas, with discernable irregular distribution of storm flood locations. The spatial distribution of annual precipitation in the Dadu River watershed is detailed in Fig. 2.1.

The complexity of hydrometeorology in the Dadu River watershed brings great challenges to hydrometeorological forecasting, flood prevention and disaster reduction, and for beneficial dispatch. The scientific regulation of flood waters, in particular, has a direct relationship to flood control safety and the economic development of Aba Prefecture, Ganzi Prefecture, Ya'an City, Leshan City, and other places in Sichuan Province, and even impacts flood control effectiveness on the middle and lower reaches of the Yangtze River. The operation management of the watershed needs to not only consider the problem of flood prevention and disaster reduction for the whole watershed, but also to consider the goal of maximizing economic benefit

© The Author(s) 2025
Y. Tu, *Management of Hydropower Enterprises*, Water Resources Development and Management, https://doi.org/10.1007/978-981-97-5584-4_2

Fig. 2.1 Distribution of
precipitation across the Dadu
River watershed (from
Spatial and Temporal
Distribution of Precipitation
in Dadu River Watershed)

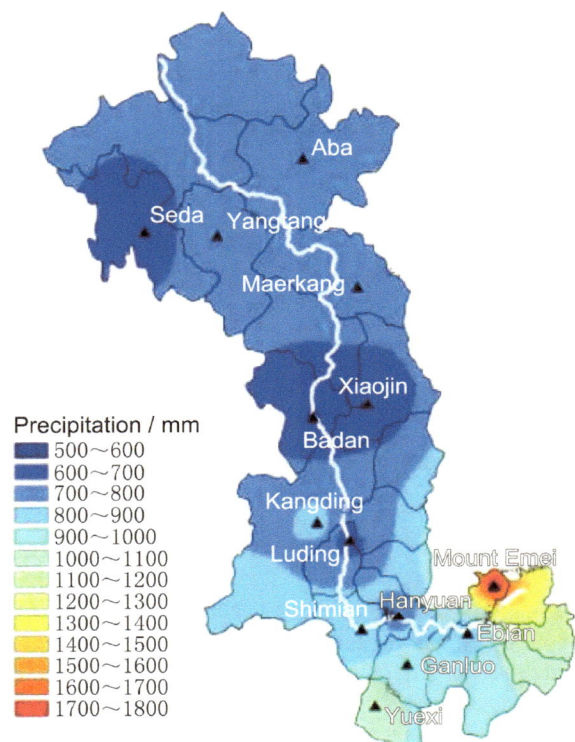

through the optimal dispatch of water resources. Therefore, how to scientifically balance safety and economics under such complex hydrometeorological circumstances has become an important subject in the operation and management of the power station group in the watershed.

2.1.2 High Occurrence of Earthquakes and Geological Hazards

The Dadu River watershed is located in the transition zone between the Qinghai-Tibet Plateau and the Sichuan Basin, falling in a region impacted by three major fault zones, the Xianshui River, the Anning River, and the Longmen Mountain faults (see Fig. 2.2), a region with complex geological conditions and a high incidence of seismogeological hazards. According to incomplete statistics, there have been 137 earthquakes of magnitude 6–7, of which 38 were earthquakes of magnitude 7 or above, including the 2008 "5.12" Wenchuan magnitude 8.0 earthquake, the 2013 "4.20" Lushan magnitude 7.0 earthquake, and the 2017 Jiuzhaigou magnitude 7.0

earthquake. The frequent earthquakes have caused high geological fragility along the path of the watershed, making the region even more prone to geological hazards, often leading to road interruptions, river blockages, equipment and infrastructure damage, and even human casualties.

In October, 2014, the results of a geological disaster survey of key zones of the Dadu River, carried out by the Chengdu Center of the China Geological Survey, showed that there exist 2521 locations with potential geological hazards along the mainstem of the Dadu River, including 785 mudslide sites, 913 landslide sites, 430 soil collapse sites, and 393 potential unstable slopes. These concealed potential seismogeological hazards seriously threaten public safety along the watershed. If any event actually did occur, it would endanger the lives and property of surrounding people, and directly threaten the personal safety of personnel in the watershed, along with the safety of power station equipment and facilities. How to effectively guard against and minimize the loss of life and property caused by seismogeological hazards

Fig. 2.2 Locations of three major fault lines

is a long-term yet urgent task facing the operation and management of the power station group in the Dadu River watershed.

2.1.3 Fragility of Ecological Environment

The terrain and geomorphology of the Dadu River watershed are complex, the climatic conditions unique, and the geological structure unstable. Vegetation types on the upper river are limited, and the degree of coverage is very low. Rocks are largely snow covered or bare. The biomass of the various natural ecosystems is limited, the system structures are simple, and the balance is poor. The ecological environment of the watershed is very fragile, highly susceptible to damage, and does not recover easily. Because of this, the Dadu River Company put forward the environmental protection concept of "partner to the green mountains and emerald waters, make the green mountains and emerald waters more beautiful." The Dadu River Company takes ecological environmental protection and governance of the watershed as an important mandate in the operation and management of the power station group in the Dadu River watershed.

2.1.4 Variety of Dam and Powerplant Types

Twenty-eight cascade power stations are planned for the mainstem of the Dadu River watershed. The dam types cover a range of gravity dams, gravel-soil core rockfill dams, faced rockfill dams, gravel-soil core rockfill dams, double-curvature arch dams, etc., with the dams high and reservoirs large, which, coupled with the complex geological structure of the river watershed and poor stability of shore slopes, only increases safety risks for the dams. If any safety problem were to occur at a dam, this might seriously affect public safety along the Dadu watershed, and even along the Changjiang River watershed. Therefore, how to safely, efficiently, and reliably control the safety of numerous hydropower dams and their reservoir shorelines is another major issue facing the operation and management of power station groups within the watershed.

The Pubugou rockfill dam is shown in Fig. 2.3.

2.1.5 Large Disparity in Equipment

Development of hydropower in the Dadu River watershed occurred over a large time span. In 1971, the first generator set of Gongzui hydropower station was put into operation to generate electricity. Subsequently, over the following half-century,

Fig. 2.3 International Milestone Project: The 186 m tall Pubugou gravel-soil core rockfill dam

additional power stations were gradually put into operation, with the current operating life of equipment reaching more than 50 years. Between old and new power station equipment, there is a large disparity in terms of design concept, manufacturing technology, workmanship, and installation quality. As the equipment in older power stations continues to age with the passing of days, the reliability drops, and the challenges in keeping the equipment safely and regularly functioning increases with each year. At the same time, development conditions for each cascade power station are different, the types of main and auxiliary equipment are different, and there is a relatively large disparity in performance parameters. Operation and maintenance methods are unalike in each case, and the job of equipment operations management for the power station group is arduous.

2.1.6 Rapidly Transforming Market

In 2015, China launched a new round of power system reforms, and the unlinking of the constituent elements of "generation, transmission, distribution, and sale" was accelerated. In 2017, Sichuan Province started to put into effect a spot market for electricity, as one of eight initial pilot provinces. Unlike other provinces, hydropower in Sichuan accounts for 80% of the total, and hydropower comprises the main body of trade in the power market. But hydropower is characterized by uncertain inflows of water, strong upstream and downstream correlation, and low variable cost in power generation, leading hydropower enterprises to face a series of new challenges. How can hydropower enterprises improve the precision of water inflow forecasting and satisfy the needs of power market trade? How can hydropower enterprises ensure the matching of cascade power generation loads, under market conditions, without increasing cascade spillage losses? In the midst of market competition, how can hydropower enterprises avoid irrational malicious competition and formulate bid offer strategies in a scientific and reasonable manner? Hydropower enterprises must actively adapt to transformations in the power market, build robust abilities to perceive the market, accurately anticipate market trends, and scientifically formulate offering strategies, so that enterprises have forward-looking discernment, the capability to analyze the overall situation, and a capacity for scientific decision-making. Only then will they be able to respond to changes in the external environment with agility and ensure the enterprise's continuing operating effectiveness.

2.2 The Challenges Facing Enterprise Management

2.2.1 People's Needs Have Become More Diverse

For traditional power plants, tasks like operational maintenance, operation dispatch, equipment servicing, and administrative management all need to be handled on site. However, hydropower stations on the Dadu River are regularly located in the depths of the mountains, far from socially developed front-line cities. Large numbers of hydropower plant employees must continue at their work locations for long periods, far from the cities and far from their families. Living conditions are difficult. With the modern improvement of living standards, employees have higher and higher expectations for improved working conditions, their individualized and diverse needs are increasing, and their aspirations for a comfortable working environment only become more intense every day. At the same time, people's pursuit of work value in their job positions is becoming more and more insistent. Modern social development has created more and better opportunities and conditions for people, and it is difficult to simply rely on traditional ideological and political work, along the old administrative management methods, and still expect the same effect as in the past.

Because of this, enterprises must tie together the transformation of ways of working with the improvement of employee working conditions, liberating employees from harsh work environments, mechanical repetition, and hard and heavy work, and satisfying their life pursuit of happiness and their pioneering and forging work spirit. Only this is a way to a fundamental solution.

2.2.2 The Connotation of "Things" Has Become Richer

Things and people are the two basic elements of enterprise management. With the continuous innovation of technology, the connotation of things is also changing. In regard to the management of the Dadu River watershed, change in things is mainly reflected in two aspects: First, in terms of technological performance, the level of informatization of equipment is constantly increasing. The original simplistic model of manually making on-site inspection rounds, along with traditional operational methods and backwards control measures, is no longer workable. This means that many functions end up being limited, and technical efficacy is very low, or can never come into play at all. Second, in terms of coordinated system operation, "things" are gradually progressing from solitary mechanical bodies, towards smart bodies represented by machine groups and digital systems. For these smart bodies, automatization and digitization are highly integrated, such that enterprises have higher and higher need for coordinated system operation. Yet, there is a natural physical separation between the individual hydropower stations across the remote areas of the watershed, each of which has developed a large number of application systems that are similar, but have different standards. With the long-term independent operation of each system, the accumulation of large amounts of non-ordered data even further aggravates the fragmentation and information isolation between the systems, resulting in serious information gaps cross-system, cross-specialty, and cross-level within the enterprise, and the need for system coordination and interconnection is even more glaring.

It is easy to see, if the enterprise cannot correctly recognize the change in the connotation of "thing," and make corresponding transformations at the management level, then the overall development of the enterprise will long be a cycle of contradictions and constraints.

2.2.3 The Relationship Between People and Things Has Changed

Throughout the process of the three industrial revolutions of the steam age, the electrical age, and the information age, people are the subject, things are objects, and people continually hold absolute power over things. But entering the era of artificial intelligence, the subject–object relationship between people and things has changed. For example, in the past, hydropower stations primarily relied on people discovering equipment problems then subsequently solving them, thus ensuring the stable operation of the power station. But on the chance the person lacks the necessary intellectual capacity or has a weak sense of responsibility toward their work, this might create immeasurable hidden risk for the operation of the power station. With the emergence of smart wearable devices and various types of high-precision sensing equipment, "things" began to help people make up for the blind spots of on-site manual inspection, automatically warning and controlling unsafe behavior of personnel. At this time, things and people influence each other, cooperating to correct deviation, and to jointly maintain the stability and safety of the power station. The position of things has started to evolve from a pure object of social development, to a relationship with people, where things and people are both mutually subjects and objects. Enterprise management cannot remain halted at the traditional stage of "people are the subjects, things are objects," and it is necessary to keep up to adapt to the changes to this type of relationship.

Chapter 3
Planning for Intelligent Operation and Management

3.1 Overall Approach to Intelligent Operation and Management

The Dadu River Company deeply merges traditional hydropower operations management together with advanced information technologies such as cloud computing, big-data, IoT, mobile networking, and artificial intelligence, striving to create a new model of intelligent operation and management for the watershed around data-driven management, and human–computer interaction and cooperation. First, modern sensing technology is used to establish a quantitative perception system to achieve automated perception of hydrometeorological conditions, equipment status, dam status, geologic side slopes, ecological environment, etc., to accomplish multi-factor business quantification based on the core area of the watershed. Secondly, efficient transmission of data is realized through the construction of 4G/5G, Wi-Fi, and other terminal transmission networks, along with a backbone optical fiber network composed of optical fiber composite low-voltage cable plus dedicated optical cable. Thirdly, centralized integration of multi-source data is realized through establishment of a watershed-scale big-data center. And finally, we give play to the creativity of employees, smartly and efficiently mine data resources, create an algorithm model library, and construct a unified platform for intelligent operation and management for the watershed, around a core of multi-business smart cooperation, that combines as a single entity such items as optimized cascade dispatch, power station operation, equipment maintenance and servicing, dam and reservoir shoreline safety control, and ecological environmental protection. Simultaneously, changes to the traditional model of hydropower enterprise management lead to the realization of automated risk identification, smart decision-making management, and autonomous remediation escalation in the operation and management of the watershed, ensuring the watershed's safe and efficient functioning, and continuing improvements in effectiveness.

The overall approach to intelligent operation and management is shown in Fig. 3.1.

© The Author(s) 2025

Y. Tu, *Management of Hydropower Enterprises*, Water Resources Development and Management, https://doi.org/10.1007/978-981-97-5584-4_3

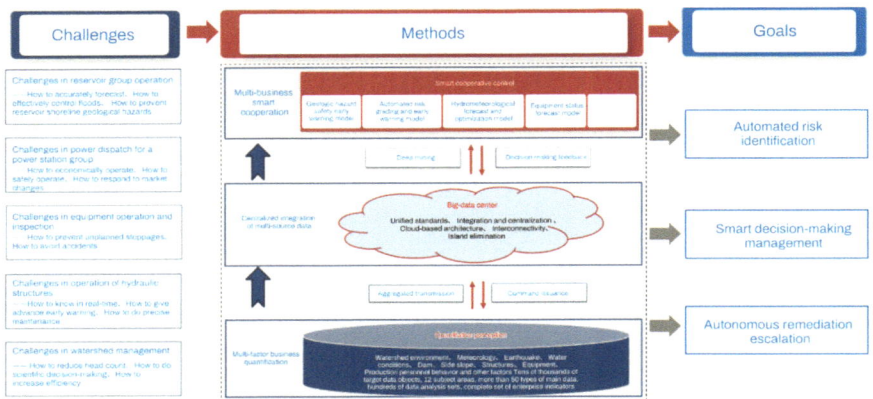

Fig. 3.1 Overall approach to intelligent operation and management

3.1.1 Multi-factor Business Quantification

Multi-factor business quantification is the foundation for realizing watershed intelligent operation and management. Through the application of various latest technologies, the Dadu River Company distinguishes perception factors for business objects, environments, and processes, so that the various businesses that the enterprise cares about can realize digitalization, transforming from the qualitative and experience-based management of the past, to a more accurate data description and data-driven management.

By the end of 2020, the Dadu River Company had established and gradually improved a set of perception systems based on the business characteristics of the Dadu River watershed, and had developed and applied IoT perception technology for core business elements. A total of 22,948 dam and reservoir shoreline side slope risk sensing points had been established, along with 110 hydrological telemetry points, and 350 types of data acquisition library for hundreds of thousands of monitoring points. Thus, the business quantification and all-round dynamic perception of major production factors such as watershed equipment operation, equipment maintenance, cascade dispatch, and environmental monitoring were gradually realized.

3.1.2 Integration of Multi-source Data

The integration and centralization of multi-source data is a prerequisite for the realization of watershed intelligent operations and management. After going through all around quantitative perception, the various factors of the entire watershed together make up a vast amount of heterogenous multi-source data. In order to achieve standardized centralized control and interconnectivity for these heterogeneous multi-source data, the Dadu River Company has built a unified integrated-deployment big-data center utilizing advanced technologies such as 5G, cloud storage, cloud computing, and big-data. The big-data center was built for the needs of the watershed, achieving unified and standardized management of storage, control, and application of multi-source data for the whole watershed, accomplishing standardized development management for application systems, avoiding stand-alone establishment of business systems, and circumventing many application stacks, effectively eliminating data islands and information fragmentation.

The architecture of the Dadu River Company's big-data center's data governance and analysis platform is shown in Fig. 3.2.

The Dadu River Company's big-data center is the base platform for achieving centralized integration of multi-source data in the watershed. Through data collection, data governance, and data access services, the on-demand flexible distribution of information resources for the whole watershed is accomplished. In the vertical dimension, this realizes smooth movement of leadership decision-making data from the base level units—functional departments—and in the horizontal dimension, this realizes data sharing across specialties and businesses, thus making enterprise application deployment more efficient, data storage more secure, data exchange more rapid, and resource utilization more effective, providing assurance support for multi-business smart cooperation.

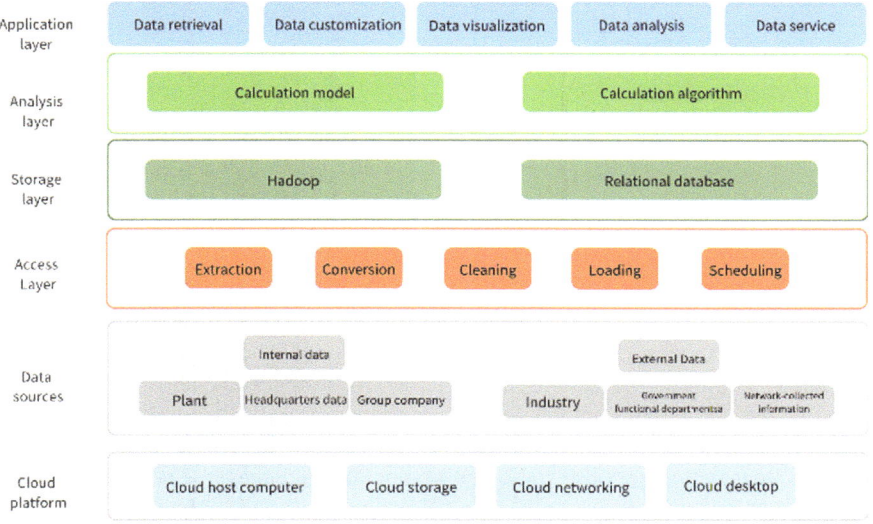

Fig. 3.2 Architecture of the Dadu River Company's big-data center's data governance and analysis platform

3.1.3 Multi-objective Smart Cooperation

Multi-objective smart cooperation is the key to realizing watershed intelligent operation and management. Revolving around various business control objectives, the Dadu River Company mines and develops enterprise big-data formed during integration and centralization, creating various smart application models, and forming a "cloud brain" for automatic risk identification, smart decision-making management, and smart cooperative control.

The production operation and management of hydropower enterprises is a systems project, which needs to comprehensively consider hydro-meteorological conditions, equipment operation and inspection, the power market, and cascade dispatch. These areas must be efficiently coordinated and attain unified objectives. In order to accelerate the realization of multi-objective smart cooperation for the Dadu River watershed, the Dadu River Company holds a big-data modeling competition every year. All employees are encouraged to combine their own operational collective wisdom and innovation, mining data at the big-data center. Employees create hundreds of different algorithmic models covering engineering construction, power production, and enterprise management. This process gradually realizes watershed intelligent operation and management of areas such as accurate forecast of hydrometeorological conditions, scientific assessment of equipment status, scientific dispatch for cascade power stations, timely early warning of geological hazards, and comprehensive assessment of dam safety.

3.2 Intelligent Operation and Management System Architecture

The Dadu River Company is responsible for the operation and management of multiple power stations in the watershed, which occurs on three levels: management of single power stations, management of power station groups, and management of the complete watershed enterprise. For this, an architecture for an intelligent operation and management system was designed, shown in Fig. 3.3.

The combination of smart systems and factors related to each enterprise level forms a system architecture of the virtual and the physical in combination, with smart autonomy, and human–machine cooperation. This architecture is capable of automated anticipation, autonomous decision-making, and self-evolution. These capabilities have different expression at each of the three levels of single power station operation management, power station group operations management, and watershed enterprise management.

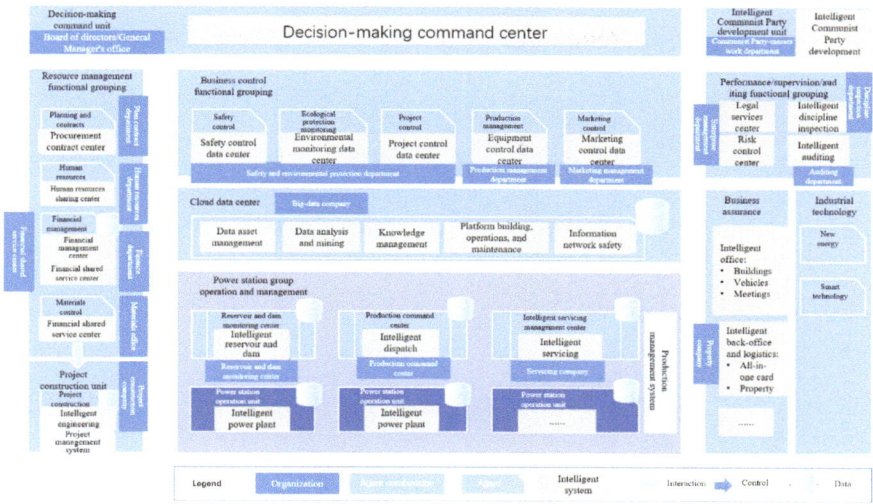

Fig. 3.3 Architecture for an intelligent operation and management system

3.2.1 Single Power Station Smart Autonomous Operation

The Dadu River Company has constructed a watershed-wide system for big-perception, big-transmission, and big-storage. Relying on its own big-computing and big-analysis capability, the Dadu River Company has formulated a smart autonomous operation system, and designed a smart autonomous power station with independent operation, smart inspection, remote control operation, and local emergency response.

- Smart Autonomous Operation System Planning

The Dadu River hydropower smart autonomous operation system has two key properties: On the one hand, power stations rely locally on on-site application of new systems and new equipment, such as comprehensive data platforms, smart inspection robots, smart safety helmets, smart safety locks, etc., to realize the independent operation of the power station, smart inspection, and on-site emergency response. At the same time, to realize remote control operation from a distance, power plants rely on watershed-level production control centers, such as the production command center, the reservoir and dam safety center, the equipment control center, and the safety risk control center. The outline for a smart autonomous operation system for a single power station is shown in Fig. 3.4.

The operation modes for smart autonomous operation of hydropower stations:

Under normal operating conditions: Autonomous power plant operation, smart inspection and remote control operation, improved level of autonomous operation, and reduced dependence on on-site personnel during the operation process are accomplished based on conventional industrial control systems such as computer monitoring systems and protection systems, and smart applications in combination, such

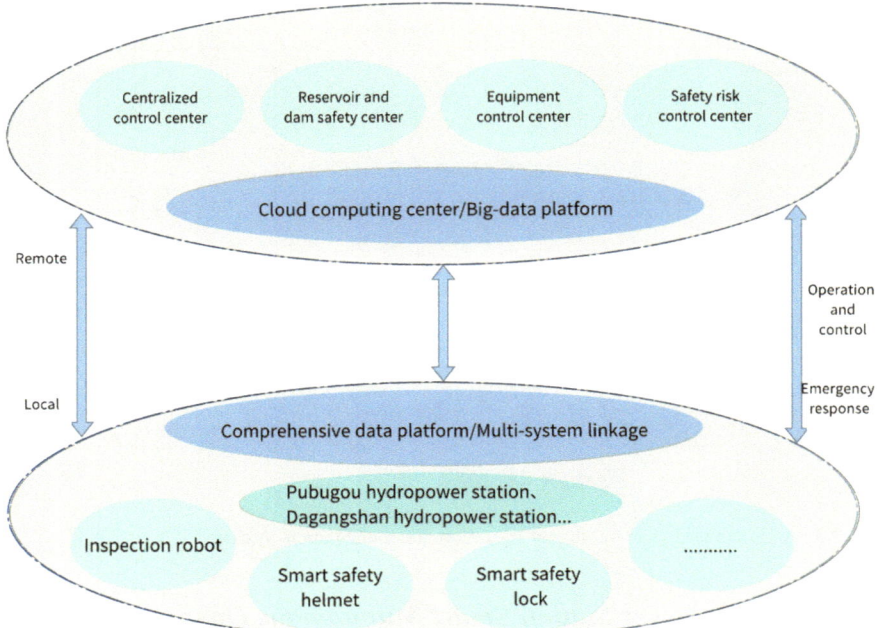

Fig. 3.4 Smart autonomous operation system for a single power station

as the power station comprehensive data platform's multi-system linkage module, the smart inspection system, and the production command center "one-button dispatch" smart regulation system.

Under abnormal operating conditions: The equipment control center accomplishes the comprehensive perception of equipment information through smart inspection and on-line monitoring systems, and applies smart analysis technology to achieve equipment health status forecasting and early warning. The safety risk control center accomplishes perception of human behavior through utilization of smart hard hats, smart safety locks, and industrial video surveillance, and warns and intervenes in unsafe human actions. The watershed reservoir and dam safety center carries out timely analysis and early warning of risk areas such as dams and shore slopes through reservoir, dam, and geological hazard monitoring systems. The on-site emergency response team, along with the production command center, carry out emergency response in a timely manner based on early warning information and emergency contingency planning.

- Smart Autonomous Power Station Design

The smart autonomous power station is designed on the basis of autonomous operation, smart inspection, remote control operation, and on-site emergency response.

Autonomous Operation: Construct a plant-level comprehensive data platform, eliminating data barriers between power station production systems, and realizing multi-system data interconnectivity. Compose a power station multi-system linkage control strategy, develop data- and message-driven multi-system linkage functions, and, while ensuring safety, break down system boundaries. Employ multi-system smart linkages of computer monitoring systems, ventilation control systems, fire protection systems, industrial television, access control security, production management systems, etc., to achieve autonomous operation of equipment under normal working conditions.

Smart Inspection: Develop smart inspection robots suitable for hydropower station application scenarios which can replace human manual inspection, along with smart safety helmets, smart keys, and other smart equipment having human–computer interaction functionality, along with smart monitoring systems for reservoir and dam safety that can replace human manual observation, thus comprehensively perceiving the status of personnel, equipment, and the environment. Using the safety risk control center, equipment control center, and reservoir and dam safety center, employ big-data mining technologies such as image recognition, voiceprint recognition, and temperature field reconstruction, to build a smart analysis model for accurate identification of human rule-breaking behavior, accurate prediction and early warning of equipment defects, and timely anticipation of environmental risks, so as to ensure the safe work of power station personnel and the safe operation of equipment and facilities.

Remote Control Operation: Construct a watershed big-transmission network. Apply technologies in communications, control, and big-data mining, etc., and using the river watershed production command center, develop a smart regulation system for cascade hydropower stations. Build a smart control model for remotely located power stations with one-button generator start-stop, one-button load adjustment, one-button switching operations, and one-button flood regulation, changing the on-site operation and control mode for power stations, and achieving on-site staffless shifts and unmanned operations.

On-site Emergency Response: Develop an on-site emergency decision-making support system, eliminate the traditional hydropower station's on-site central control room, set up an emergency command post, and prepare an emergency response team on-site or at a personnel housing area, responsible for power station on-site fault screening, anomaly handling, and emergency operation, so as to achieve efficient emergency handling on-site, and ensure the safety of the personnel, equipment, and facilities of the hydropower station under abnormal operating conditions.

3.2.2 Cascade Power Station Group Intelligent Operation

With smart autonomous operation of single power stations as a foundation, the operation of cascade power station groups in the watershed must also consider joint dispatch of cascade water resources, cooperative dispatch for power generation, and smart operation and inspection of equipment. Following advancements in technology, the Dadu River Company has realized the intelligent operation of tens of millions of kilowatts of installed capacity of multiple power stations in the watershed through the building of various platforms such as platforms for intelligent reservoirs and dams, intelligent dispatch, and intelligent servicing, thus forming an intelligent operation mode with the properties of remote centralized control, remote monitoring, graded early warning, and emergency mobility.

(1) Intelligent Cascade Reservoir Group Operation

A hydropower reservoir generally performs a number of tasks, such as flood control, power generation, ecological protection, and shipping. Systems in the Dadu River watershed comprehensively perceive factors such as those related to meteorology, water conditions, equipment, and geological hazards, accurately forecasting precipitation and runoff processes, and combining flood control restrictions, ecological constraints, shipping needs, and shore slope safety, to achieve multi-objective cooperative optimization.

The building of intelligent operation for watershed cascade reservoirs is detailed in Chap. 4.

(2) Cascade Power Station Group Intelligent Power Dispatch

Power dispatching for a cascade hydropower station group not only needs to satisfy overall plans for short- and long-term effectiveness, but also must achieve the overall plans for both the effectiveness of a single station and the effectiveness of the whole. In addition to conventional medium- and long-term optimized dispatch, flood water is utilized as a resource through technologies such as late flood season staged water impoundment. Within the context of the spot trading market, it is necessary to fully perceive the market environment, scientifically analyze the price patterns, accurately predict the market clearing price (MCP), and formulate a rational offering strategy. In terms of real-time dispatching, safe, economical, efficient, and intelligent power generation operations in the cascade are achieved through real-time cooperative cascade dispatch combining medium- and long-term optimized dispatch objectives and market trade results.

The building of intelligent cascade power generation operations is detailed in Chap. 5.

• Intelligent Operation and Inspection of Watershed Equipment

Operation and inspection, in terms of hydropower station equipment in the Dadu River watershed, include the operation, maintenance, and servicing of hydropower station equipment. Equipment operation includes carrying out tasks related to equipment such as operation control, status monitoring, and emergency handling. Equipment maintenance includes carrying out tasks such as daily equipment maintenance, defect handling, fault analysis, fault clearance, and other tasks to keep equipment in good technical condition. Equipment servicing includes analyzing and predicting the health of equipment, and repairing equipment having deteriorated performance or failure, in order to restore equipment functionality and improve technical condition.

Technologies such as big-data, artificial intelligence, and knowledge management are used on the Dadu River watershed to accomplish the cooperative blending of human actions with smart equipment and systems across the areas of equipment operation, status monitoring, emergency handling, and fault analysis, thus achieving the intelligent operation and inspection of equipment.

The building of intelligent operation and inspection of watershed equipment is detailed in Chap. 6.

- Intelligent Operation of Hydraulic Structures

Dams, water diversion, and flood discharge facilities are important hydraulic structures for hydropower stations, and play an important role in flood control, power generation, irrigation, and shipping. Through advanced perception technology, the Dadu River Company accurately monitors trends of external and internal change in hydropower station hydraulic structures, and constructs a risk assessment model, providing decision-making support for scientifically formulating operation modes and handling risk, and effectively bringing major risks in the process of operation and management of hydraulic structures under control.

The building of intelligent operation of hydraulic structures in the watershed is detailed in Chap. 7.

- Intelligent Ecological Environmental Protection

Given the practical considerations related to intelligent power production by power station group joint dispatch, the Dadu River Company dynamically acquires ecological environmental protection data such as water environment, acoustic environment, light environment, atmospheric environment, aquatic organisms, and soil erosion, through multi-dimensional, high-precision, and automated collection. At the same time, intelligent dynamic adjustment is carried out in the power plant area directed at indicators such as noise, air, solid waste, water environment, ecological flow, habitat, aquatic ecology, terrestrial ecology, and other indicators. The method of reservoir dispatch is improved, the operation mode of fish passage facilities is scientifically adjusted, and the fluctuation in survival parameters such as water temperature, dissolved oxygen, and pH value, are reduced. Thus, intelligent control capabilities are realized in areas such as automated environmental risk identification, river health assessment, graded early warning, and assisted decision-making support.

The building of watershed intelligent ecological environmental protection is detailed in Chap. 8.

3.2.3 Watershed Intelligent Enterprise Management

Dadu River Company is the primary enterprise operating and managing the watershed power station group. The company has established a matching intelligent enterprise management model at the levels of strategic control, business control, resource allocation, metric supervision, business assurance, and has engaged in effective business transformation and management innovation.

1. Strategic Control

As the principal operation and management unit for the watershed power station group, the Dadu River Company bears main responsibility for watershed environmental protection, reservoir shoreline safety control, flood and high-water control, and water resources dispatch. The Dadu River company established intelligent units (unit brains) in the base level operational units, and specialty data centers (specialty brains) in the company offices, and simultaneously built a decision-making command center, which effectively improved strategic mastery and risk control capabilities.

*   Decision-making command center: With the goal of "smart control of major risks, control of major business processes, and smart support of major decisions," by applying new technologies such as big-data and artificial intelligence, and by bringing together specialty brain and unit brain data for the complete watershed, establish cross-specialty smart early warning models to achieve smart early warning of major risks, smart analysis of risk causes, and decision-making support for emergency events.

2. Business Control

The Dadu River Company has built a number of data-driven, specialty-strengthening, vertically connected, and horizontally cooperative business specialty data centers (specialty brains) in its headquarters. Highly reliant on the cloud data centers, these specialty data centers focus on intelligent construction of business controls such as project control, production management, safety management, environmental protection, and marketing.

*   Project control data center: With the goal of "project full-element control and project full life-cycle management," founded on an intelligent project-risk early warning and control system along with control modelling, supported by enterprise big-data, on the basis of smart daily management of the watershed project, with focus on management of the major issues of safety, quality, progress, investment, environmental protection and other key parts of the project, through big-data and decision analysis models, achieve analysis, early warning, decision-making, and comprehensive management of the trends and systemic problems in project management elements.
*   Production control data center: Guided by equipment full life-cycle management, following the approach of comprehensive monitoring, quantitative assessment, smart analysis, and scientific arrangement, comprehensively command the status and operational performance of equipment assets, rationally deal with equipment anomalies, defects and accidents, scientifically arrange equipment maintenance and servicing, and ensure the reliability, safe operation, and rational servicing of equipment.

- Safety control data center: With the goal of "smart object monitoring, automated risk assessment, and centralized scheduling of resources," through smart equipment, comprehensively perceive the safety status of power production units, and through the establishment of a comprehensive safety risk control system, achieve daily real-time monitoring of safety, advance anticipation of safety anomalies, timely response to safety incidents, and iterative improvement of safety control.
- Sales control data center: With the goal of "scientific marketing and all-around expansion," break the traditional marketing model, comprehensively perceive the market environment, establish marketing big-data, accurately anticipate market element change trends, scientifically formulate marketing strategies, optimize marketing programs, and enhance electrical power marketing capabilities.

3. Resource Allocation

Efficient allocation of resources is a core management task for enterprise management. Employing the big-data center, the Dadu River Company has improved the company's resource allocation and management capabilities by building four types of specialty data center (specialty brains).

- Plan contract center: With budgets, plans, and contracts as the main theme, achieve the overall planned arrangement, rational allocation, and efficient running of production operation tasks and resources. Through process monitoring and the strengthening of specialties, along with functions such as early warning and cross-specialty cooperative analysis, effectively improve management efficacy, improve resource use efficiency, and create greater operational effectiveness.
- Human resources shared service center: With the goal of "serving employees, optimizing structure, and upgrading the organization," realize the comprehensive management of human resources, and through analytical modelling, optimize personnel structure, and enhance the value of personnel, satisfying the ever-changing human resource needs of the new environment.
- Financial shared services center: With the goal of "business-finance integration, smart control, and flexible sharing," achieve the effective integration of financial resources and improve efficiency, while simultaneously strengthening control, controlling risks, and enhancing the value position of financial management within the enterprise.
- Material control center: With the goal of "centralized control, data-driven, risk early warning, and scientific decision-making," achieve whole-process management of material demand, planning, procurement, distribution, and inventory management.

4. Metric Supervision

An effective system for assessment, supervision, and auditing of metrics is an assurance mechanism for compliant enterprise operation. In the past, compliance control, discipline inspection, and auditing, all relied on manual operation, with long cycles, low efficiency, and limited means. With the Dadu River Company's full digitization at the production and operations levels, and in reliance on the company's cloud data center, through the creation of specialty brains such as comprehensive risk management center, intelligent discipline inspection, and intelligent audit, build novel data-driven measurement, supervision, and audit models, achieving faster, more efficient, and more accurate management efficacy, producing noticeable management results.

- Legal and risk control center (compliance control center): With the goal of "comprehensive monitoring, graded control, and cooperative response," with the risk management system as a foundation as well as system assurance based on internal control/performance, build an organizational culture of risk control across the enterprise, and by carrying out real-time monitoring of risks in the whole enterprise, realize graded early warning and control of risk, while simultaneously situating comprehensive legal services in its proper organizational placement, to professionally guard against legal risks.
- Intelligent discipline inspection data center: With the goal of "ubiquitous supervision and smart assessment," and sustained by the IT environment, build a one-stop on-line discipline enforcement platform, realizing real-time supervision of all discipline inspection and supervision risk points in the whole enterprise, and achieve specialized and smart support for discipline inspection supervision, not only maximizing the prevention of violations of discipline and rule-breaking behavior, and ensuring the total implementation of laws and regulations in the enterprise, but also allowing employees to actively participate in the discipline inspection and supervision process, safeguarding the rights of employees and the interests of the enterprise.
- Intelligent audit data center: With the goal of "centralized management, comprehensive coverage, division of labor and cooperation, and responsiveness," utilizing the unitary audit platform and audit big-data center as a foundation, achieve interconnectivity between the audit work and the business under audit, and raise the level of automation of audit work, and strengthen the validity of the audit results.

5. Business Assurance

Business assurance primarily includes support for the daily administrative office work as well as back-office and logistics support services work. With the comprehensive application of digital technology as foundation, the Dadu River Company has achieved an efficient, rapid, and highly humanized experience for office work and back-office and logistics services through the construction of an intelligent office and intelligent back-office and logistics platform.

- Intelligent office: Through an office service center, achieve centralized management of files and official documents, meetings, electronic seals, external activities, as well as approvals and processing, improving office efficiency.
- Intelligent back-office and logistics: Realize centralized management of corporate buildings, properties, vehicles, safety/security, reception services, etc., and raise employee sense of happiness.

3.3 The Path to Intelligent Operation and Management

Digital technology is an important tool to drive the adoption of intelligent operation and management for the watershed. In order to effectively achieve the shift to watershed intelligent operation and management, the Dadu River Company formulated a "five-bigs" implementation path.

3.3.1 Forging a "Big-Perception" System

The key to building a "big-perception" system lies in the successful application of Internet of Things (IoT) technology, industrial automation technology, and the building of internal and external data acquisition and exchange platforms for the enterprise to achieve a comprehensive perception of enterprise factors. The Dadu River Company has made a large number of deployments in perception related to hydrometeorology, power plant operation, and reservoir and dam safety.

In terms of hydrometeorological perception, the company has built an automated water monitoring and reporting system consisting of 110 hydrological telemetry stations covering the whole watershed, and integrated meteorological data from the top institutions such as the National Meteorological Center of China, the National Weather Service of the United States, and the European Centre for Medium-Range Weather Forecasts. Every day, the company carries out 10 gigabytes of hydrometeorological big-data analysis, plotting the distribution of rainfall across the watershed in a grid format, to accurately grasp localized water inflow and rainfall conditions.

In terms of perception related to power plant operations and maintenance management, a total of 128 common fault indicators were defined for multiple components, such as water turbines, generator stator/rotors, transformers, and speed governors, with more than 7,000 status monitoring measurements captured. An equipment database was established, containing video, audio, infrared, vibration, and other on-line monitoring data for 350 kinds of equipment in 12 categories, so as to grasp the level of operational health of equipment in a timely manner. A defects database was instituted, focusing on defect phenomena, defect sites, defect causes, and defect handling methods, and recording more than 2,000 standard defects, to improve the accuracy and timeliness of fault pinpointing and hidden risk troubleshooting. At the same time, standardized equipment and material coding was completed for the whole watershed, resulting in more than 80,000 hydropower five-segment material codes and more than 260,000 six-segment equipment codes, which greatly raised the management level of equipment and materials. Perception capability towards on-site equipment, personnel, and environment was also tremendously expanded on the foundation of a full set of perceptive equipment, such as smart inspection robots and smart safety helmets.

Fig. 3.5 The automated monitoring and perception system of Pubugou Dam

In terms of perception of reservoir and dam safety, using reservoir and dam smart monitoring technology, 22,948 operational risk measurement points in 9 categories were established across the watershed power station group and surrounding mountains, covering factors such as earthquakes and other environmental impact factors. Risk measurement standards were formulated so as to realize cooperative smart perception of the dynamic real-time state of the mountains, the reservoir dams, and side slopes. Real-time collection and mastery of various types of data such as displacement, deformation, and settlement was accomplished. This all completely replaced the previous method of relying on manual monitoring and reporting of measurement point data, and represents a qualitative leap in terms of the completeness, real-time relevance, and effectiveness of data.

The automated monitoring and perception system of Pubugou Dam is shown in Fig. 3.5.

3.3.2 Building a "Big-Transmission" Network

The key to building a "big-transmission" network lies in building an extensive transmission network system based on the internet, supplemented by industrial IoT and mobile networking, providing an extensive "highway" for the Internet of Everything (IoE), and achieving the rapid, real-time, and massive transmission of data.

A communication transmission network covering the production area of the watershed was established, having a 10-gigabit network backbone, and gigabit-to-desktop network connections, with an optical-fiber core network, optical-fiber local access networks, a mobile private network, as well as an emergency communications satellite network. A backbone communication transmission network was built, centered around a 2.5 gigabit optical fiber composite low-voltage cable transmission network, and supplemented by a 150 Mbits commercial private line. A mobile private network covering the key reservoir area of the Dadu River mainstem was completed, and a mobile network covering the production area of the watershed was established with full 4G (and partial 5G) coverage and complete high-speed Wi-Fi coverage. A Ku-band VSAT emergency satellite communication network covering the whole watershed was established, and a KA-band satellite mobile internet configured.

3.3.3 Constructing a "Big-Storage" Platform

The key to building a "big-storage" platform is to build a big-data center based on hardware devices such as servers, storage, networking, security, and utilizing various types of virtual resource pool, to achieve massive storage and rapid extraction regardless of data shape.

Through the establishment of a big-data center, the Dadu River Company has built a unified data storage platform that includes the ten major subject areas of the watershed. The adoption of a hybrid storage mode of "centralization + distribution," which mixes traditional SAN optical fiber storage with distributed fusion storage technology, has achieved data tiering, providing 1400 TB of storage space, as well as the ability to smoothly upgrade capacity for future needs. Currently, the big-data center has centrally consolidated more than 300 TB in data resources, covering nine large hydropower stations across the whole watershed.

3.3.4 Raising the Level of "Big-Computing"

The key to improving the level of "big-computing" lies in the use of advanced algorithm technology, along with computing power technologies such as automated learning and deep learning, improving the algorithms and computing power of the cloud computing center, so as to provide support for intelligent operation and management data processing.

By strengthening the underlying hardware infrastructure and establishing a unified smart management platform, the Dadu River Company has achieved cloud-based deployment and unified management for all servers. Simultaneous adoption of a distributed architecture reduces the potential impact of single-node failure on the entire system. Huawei virtualization technology based on OpenStack architecture provides IaaS layer cloud services and shifts various enterprise big-data to the cloud, thereby building a computing resource pool with more than 1800 CPU cores and 29 TB of internal memory. With the approach of "one directory, one map, and one table," the work of data governance and data standard establishment was fully rolled out, and data interconnectivity and in-depth mining functions improved, so as to provide support for assisted decision-making and risk early warning. The architecture for virtualized server cloud computing is shown in Fig. 3.6.

By the end of 2020, the cloud computing resource pool hosted more than 350 virtual servers, providing computing and storage services to the 20 plus specialized data centers involved in the intelligent operation and management of the Dadu River watershed. At the same time, through desktop virtualization technology, the cloud desktop was fully applied to replace office computers at the company's headquarters and at many subsidiaries, so as to realize the unified establishment, unified management, and unified allocation of cloud computing pools and cloud desktop pools. Simultaneously an IT control platform was built to realize unified monitoring and management of information resources and network security, and ensure network and data security.

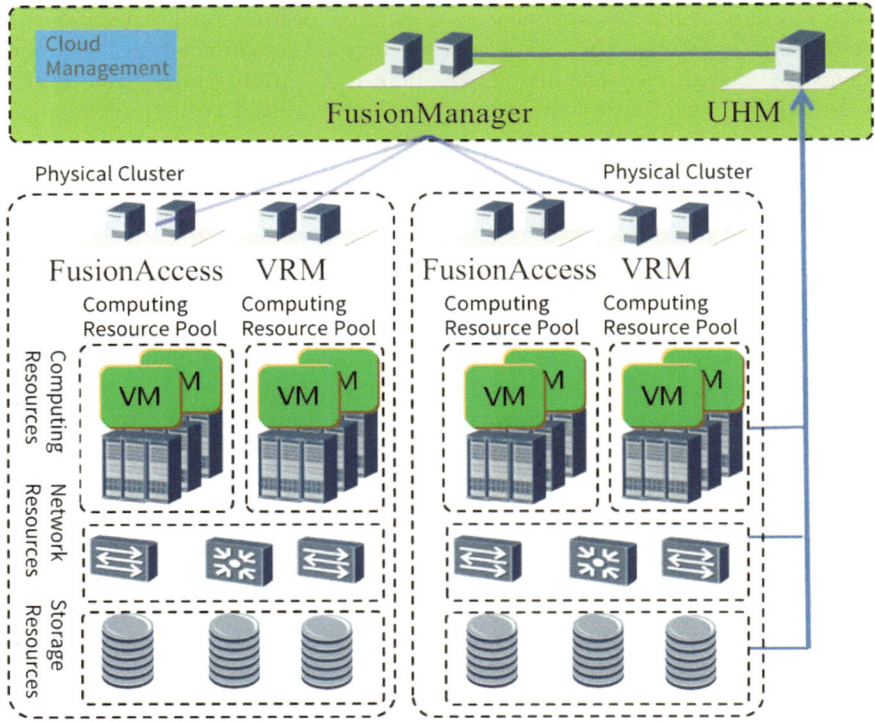

Fig. 3.6 Architecture for virtualized server cloud computing

3.3.5 Cultivating "Big-Analysis" Capability

The key to cultivating "big-analysis" capability lies in multi-dimensional analysis and mining of data, building various smart prediction, early warning, and decision-making models for intelligent operation and management of the watershed, in order to provide support for watershed power plant group operation and management that achieves automated risk identification, smart decision-making and management, and autonomous remediation escalation.

Only when big-data is actually applied to specific business scenarios can it produce real value. The Dadu River Company came up with a series of typical "big-analysis" business application scenarios through innovative means such as big-data modeling competitions, intelligent enterprise discussion salons, and youth innovation workstations. For example, in equipment fault troubleshooting, an internally developed inspection and warning robot not only has the ability to carry out analysis and processing at the front-end, but can also transmit various types of images collected at the front-end, along with information related to temperature, vibration, gases, etc., to the data backend to carry out analysis and calculation, and also push equipment operation anomaly alerts to relevant personnel, greatly improving the capability and efficiency of emergency incident analysis. In terms of equipment operational trend analysis, the patented hypersphere modeling technology machine learning engine is used to mine historical data of 29 key indicators for the generator set, establish a health status perception model for the generator set equipment, thus achieving the digital assessment of equipment health level and developing trends. In terms of security system linkage analysis (as shown in Fig. 3.7), by changing from linkages between a few systems or between scattered fragmentary systems as was previously, interconnection now exists between more than 10 subsystems, such as between power plant monitoring, excitation, and fire-fighting systems. Through a comprehensive data analysis, when emergency events occur, system smart linkages are accomplished, improving the team's emergency handling capability.

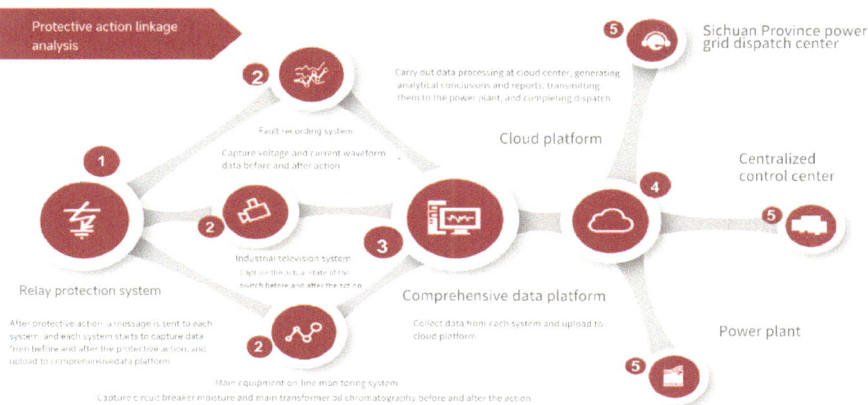

Fig. 3.7 Dadu River power station security system linkage analysis diagram

Chapter 4
Cascade Reservoir Group Intelligent Operation

4.1 Approach and Goals

The intelligent operation of the Dadu River cascade reservoir group is aimed at real-time perception, accurate prediction, and smart regulation. With automation, informatization, and digitalization as a foundation, technologies such as cloud computing, big-data, IoT, mobile internet, and artificial intelligence are applied to build an autonomous learning and smart decision-making model, achieving multi-objective smart cooperation in subjects such as flood control, safety, ecology, shipping, and water supply.

- Real-time Perception

With specialized data centers as a foundation, data related to hydrometeorology, shoreline side slopes, and equipment status is automatically perceived, with simultaneous perception of each production factor.

In terms of hydrometeorological conditions, every 5 min in real-time the company collects water and rain data from 110 self-built telemetry stations, and on an hourly rolling basis, brings in varied sets of meteorological data from US and European meteorological sources as well as from Chinese central and provincial meteorological centers, for a total of about 7,300 grid squares and 180,000 pieces of data per day, thus achieving highly centralized integration of basic hydrometeorological data.

In terms of reservoir shoreline side slope safety, tens of thousands of automated measurement points have been established to collect, identify, and interact in real-time, integrating multi-source data such as dam monitoring, water conditions, facility operating status, environment, and boundary information, to achieve real-time appraisal of reservoir and dam safety risks, and overall evaluation of operational status.

In terms of equipment information, in association with smart sectors such as "the intelligent power plant" and "intelligent servicing," equipment status, and operational information are comprehensively collected, hydropower plant equipment real-time

© The Author(s) 2025
Y. Tu, *Management of Hydropower Enterprises*, Water Resources Development
and Management, https://doi.org/10.1007/978-981-97-5584-4_4

health status and change trends of are mastered, and hydropower plant alarm information is standardized. Based on this, graded smart filtering and notification of power plant alert information is achieved, so that alerts can be rapidly and accurately judged and handled by dispatch personnel.

- Accurate Forecasting

With comprehensive data perception and interconnectivity as a foundation, big-data analysis technology is used to build multiple types of smart forecast and early warning models to achieve accurate forecasting of key factors such as meteorology, water conditions, dam deformation, and side slope displacement.

In terms of hydrometeorological forecasting, a self-built Weather Research and Forecasting (WRF) and multi-source advantage fusion forecasting model for the watershed with a resolution of 1.5 km allows meteorological elements such as rainfall and temperature to be predicted on a rolling basis, by the day, 10 days in advance, and by the hour, two days in advance, achieving a breakthrough from qualitative to quantitative forecasting. At the same time, meteorological data is coupled to hydrological forecasts, and a runoff forecasting system was built, with forecasts such as similarity forecast, Xin'an River forecast, and probability forecast operating in parallel.

In terms of reservoir shoreline safety risk early warning, based on coupling mechanisms between risk factors and reservoir shoreline catastrophe mechanisms, a cascade shoreline safety risk control system for the watershed was introduced. A smart shoreline safety risk control platform for the cascade was created, a first with such characteristics as data reliability analysis, multi-source information fusion, autonomous risk anticipation, and early warning response regulation, achieving reservoir shoreline safety risk early warning and responsive decision-making.

- Smart Regulation

Relying on real-time perception and accurate prediction, and in achievements in intelligent servicing and in the intelligent power plant, and employing artificial intelligence technology along with a core model for multi-dimensional objective cooperative optimized decision-making, the Dadu River Company has optimized the operation process and reshaped the operation mode, achieving smart "one-button dispatch" of the watershed's cascade reservoirs. By breaking the original experience-based dispatch mode, a whole dispatch chain—of accurate prediction of hydrometeorological conditions, real-time perception of reservoir shoreline safety, anticipation of flood control risks, smart decision-making dispatch solution, and automated issuing of dispatch commands—is generated by "one-button," thus realizing dispatch decision-making and response in one go.

The intelligent operation architecture for cascade reservoir operation is shown in Fig. 4.1.

By the end of 2020, the Dadu River Company had already put into operation nine reservoirs, at Houziyan, Dagangshan, Pubugou, Shenxigou, Zhentouba (Level I), Shaping (Level I), Gongzui, Tongjiezi, and Jiniu, successively connecting them

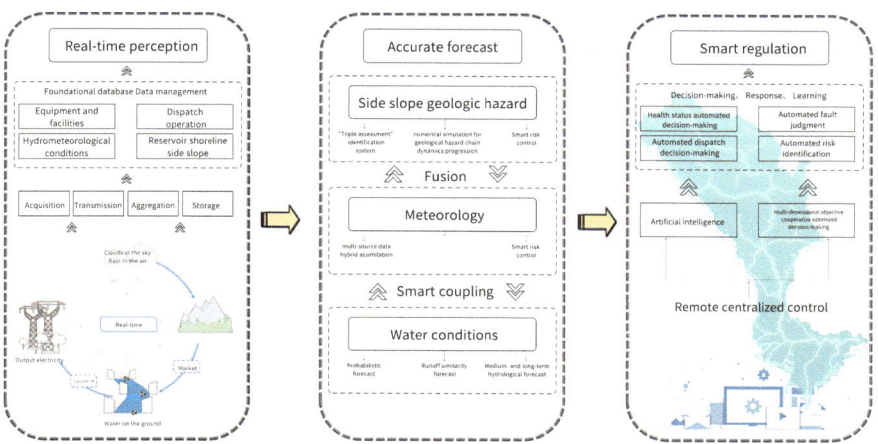

Fig. 4.1 Intelligent operation architecture for Dadu River cascade reservoir operation

to the cascade intelligent dispatch system, and realizing overall the objectives of real-time perception, accurate prediction, and smart regulation.

4.2   Key Technology

4.2.1   Precision Big-Data Hydrometeorology Forecasting Technology

Hydrological forecasting is a foundational work for reservoir dispatch operations, and forecasting effectiveness directly impacts the scientific nature of dispatch decision-making. Meteorological forecasting is the foundation for hydrological forecasting and is a key factor in extending the effective forecast period of hydrological forecasting. The low accuracy of numerical meteorological forecasting, the difficulty of quantifying the uncertainty of hydrological forecasting, and the spatio-temporal variability of hydrological processes are all common long-term challenges faced in the field of hydrology, especially in many watersheds in the southwest plateau region of China. Due to the complexity of geographical and climatic characteristics in the region, these problems are even more pronounced, and the limitations they cause have been called a "forecasting ceiling" in the industry. After years of investigation, the field of hydrology has gradually formed a relatively complete theory of hydro-logical circulation in river watersheds, realizing the full integration of meteorolog-ical and hydrological disciplines. Hydropower development enterprises in all major watersheds, and flood control authorities at all levels, are also constantly looking for hydrometeorological forecasting technologies suitable for their own particularities, in order to further raise the operations management level of reservoirs and better tap

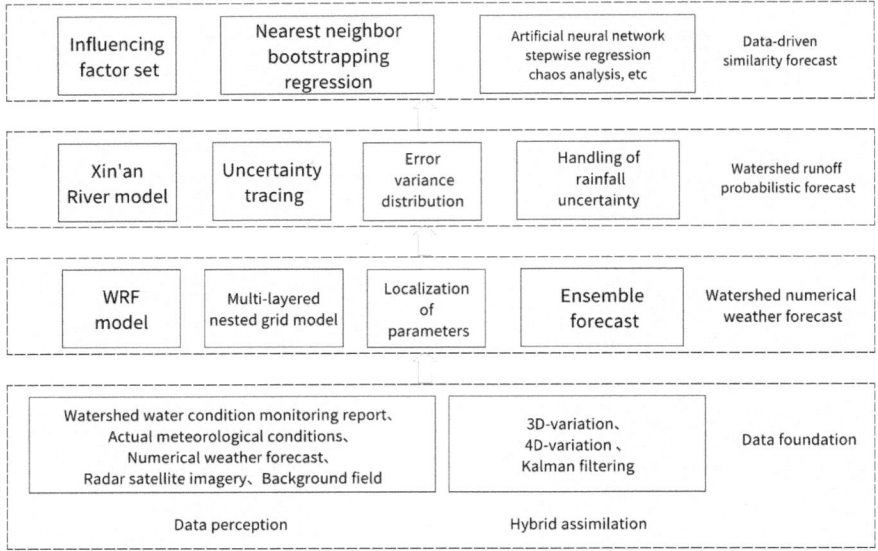

Fig. 4.2   The Dadu River hydrometeorological forecast system

the effectiveness of reservoir regulation. Through the establishment of a deeply integrated IUR (Industry-University-Research) unified research mechanism, the Dadu
River Company has continuously innovated while also introducing new technologies
and new achievements, and has gradually formed a hydrometeorological forecasting
system with its own special characteristics, suitable for the numerous high-plateau
watersheds in southwest China, shown in Fig. 4.2.

1. Building Up a Multi-source Hydrometeorological Data Foundation

- Multi-source Data Perception

A network consisting of 110 hydrometeorology telemetry stations is the basic method
to obtain real-time water and rainfall information in the Dadu River watershed.
This meteorological information is enhanced through information from specialized
organizations such as the China Meteorological Administration and other organizations, by including materials from their various stations within the watershed,
gaining precipitation, temperature, and other meteorological information in real-
time as monitored by these specialized meteorological centers. At the same time,
with developments in numerical weather forecasting technology, the Dadu River
Company pulls in gridded precipitation and air temperature factor materials from
assimilated high-altitude and satellite-derived multi-aspect measurement information, precipitation retrieval data measured by Doppler radar, and numerical weather
forecasting products from various countries, such as GRAPES from China, ECMWF
from Europe, NCEP from the United States, etc. The sources and content of input data

used in the Dadu River numerical weather forecast system can be seen in Table 4.1. On this underpinning, the introduction of global forecast field meteorological factors as background field information was to provide a foundation for the watershed's own localized numerical weather prediction model.

- Multi-source Hybrid Data Assimilation

Broad acquisition of multi-source information progressively enriches the watershed meteorological information base, but the diverse meteorological materials are not perfectly compatible in terms of spatial and temporal distribution, such as having 5 km × 5 km, or 20 km × 20 km, spatial resolution, or having minute or hourly temporal resolution. How to make full use of the multi-source observational data, sensibly converting them to the same spatial and temporal scale, is an important issue that must be solved before data application.

To this end, organically coupled 3D/4D-variational assimilation and ensemble Kalman filtering is used in the Dadu River watershed, and different hybrid assimilation solutions are implemented by zone and time period. This allows multiple types of observational material to relatively accurately represent current atmospheric conditions, further improving the quality of observational data analysis and the accuracy of numerical forecasting.

2. **WRF Model-Based High-Resolution Meteorological Forecasting Technology**

The Dadu River watershed WRF model employs a multi-layered nested grid model design (as shown in Fig. 4.3), utilizing geographic data with matching resolution to

Table 4.1 Sources and content of input data used in the Dadu River numerical weather forecast system

Data source	Data content	Retrieval frequency
Network of Dadu watershed water condition monitoring and reporting stations	Measured data for station rain volume, water level, and flow	Every 5 min
China Meteorological Administration	Measured data for station precipitation, air temperature, etc.	Hourly
	Data for gridded watershed precipitation, air temperature, etc.	Hourly
	Doppler radar precipitation retrieval data	Hourly
	Numerical weather forecasting products (China, Europe, United States, etc.)	Daily
United States National Centers for Environmental Prediction	Global grid fields	Every 6 h
The European Centre for Medium-Range Weather Forecasts	Global grid fields	Every 12 h

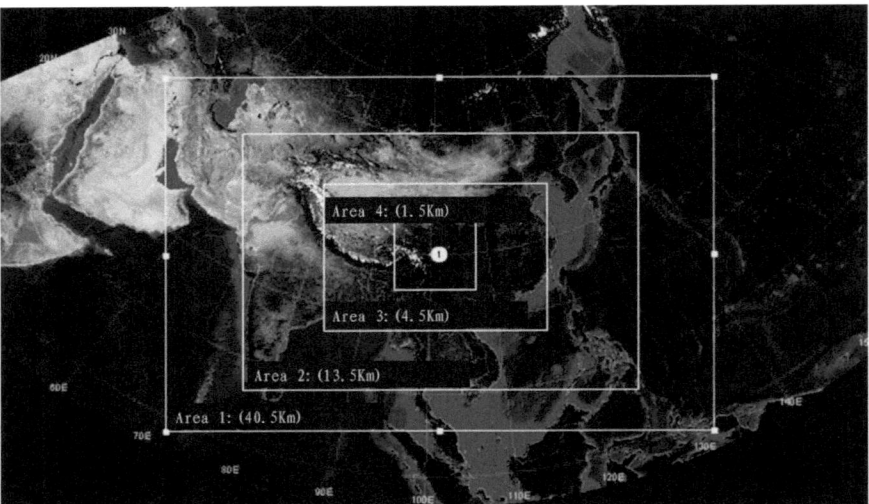

Fig. 4.3 WRF model nested region design for the Dadu River watershed

generate optimal static geographic information data fields (including terrain height, land use type, and albedo) for multiple nested zones. Among these nested zones, the innermost zone covers the watershed itself, while the outermost zone largely covers the range of weather systems that might impact the watershed.

At the time that the WRF model for the Dadu River watershed was established, in consideration of the topographic and climatic characteristics of the watershed, parameter solution combinations were established for separate zones and time periods, and cognitive computational self-adaptive parameter tuning technologies were built based on optimization-simulation and statistical models. By carrying out adaptive parameter optimization and forecast error correction, statistical counts of forecasts from previous months are conducted to analyze the relationship between numerical forecast model results and actual conditions, to search out the statistical characteristics of deviations, and summarize the various combinations of parameterized solutions with the highest forecast accuracy in previous months or even days, then applying this to future precipitation forecasting.

3. **Probabilistic Runoff Forecasting Technology Integrating Numerical Weather Prediction**

The Dadu River runoff probability forecast is an enhancement of the deterministic forecast of the Xin'an River model. By quantifying the main uncertainties in each individual constituent part, such as rainfall input uncertainty, model structure uncertainty, and model parameter uncertainty, carrying out quantitative analysis of forecast uncertainty, maximizing the utilization of all types of information in the forecast process, and describing the hydrological forecast uncertainty process both quantitatively and in the form of probability distribution, this provides, for a flood instance or runoff process, factors such as maximum possible flow rate, minimum possible flow rate, probability of occurrence of a certain flow magnitude, etc. The application of probabilistic forecasting technology not only raises forecast precision, but even more so provides enriched information elements for forecasting, and stronger support for flood-control regulation decision-making.

4. **Similarity Runoff Forecasting Technology Based on Data Mining**

In recent years, with the establishment and improvement of the modern hydrometeorological measurement network covering the whole watershed, historical hydrometeorological data are accumulated continuously. Concurrent rapid development of artificial intelligence and big-data technology opens up new approaches and new paths to break through the runoff forecasting bottleneck. With the help of big-data technology, the innumerable pieces of historical data can be analyzed comprehensively and at multiple levels, to dig out the hydrological laws hidden behind the numbers.

For a particular watershed, prevailing weather systems that condition rainfall will recur continually, and under similar weather system conditions, ensuing rainfall and runoff processes will also be similar. When there are historical rainfall and runoff materials covering a long period, data mining technology can be used to "predict the future with reference to the past," that is, to predict future runoff based on the information provided by similar historical rainfall and runoff. This is of great significance toward improving the refined management and lean dispatch of cascade reservoir groups in the watershed.

- Both Mechanism-Driven and Data-driven Short-term Runoff Similarity Forecasting

For short-term runoff forecasting, a similarity forecasting model that couples the advantages of a data-driven model and a process-driven (mechanism-driven) model was established in the Dadu River watershed (as shown in Fig. 4.4). By analyzing the rainfall-runoff physical causal processes and identifying watershed runoff forecast factors, and additionally searching for rainfall-runoff similarities through data mining techniques, future flood processes are forecast using a multi-factor nearest neighbor bootstrapping regression model.

Establishing a spatial mapping relationship from point rainfall amount to surface rainfall amount reduces the input dimensionality of rainfall data while also reflecting the spatial distribution pattern of rainfall. Sliding window sampling is utilized to build a historical rainfall-runoff sample library to solve the problem of insufficient measured data series. An overall weighted rainfall-runoff similarity measurement index scales the differences in rainfall and runoff magnitude, and parameters are optimized via smart optimization algorithms. Real-time access to rolling 7-day rainfall forecast results extends the forecast period, providing rolling prediction of the runoff process in the watershed for the coming week.

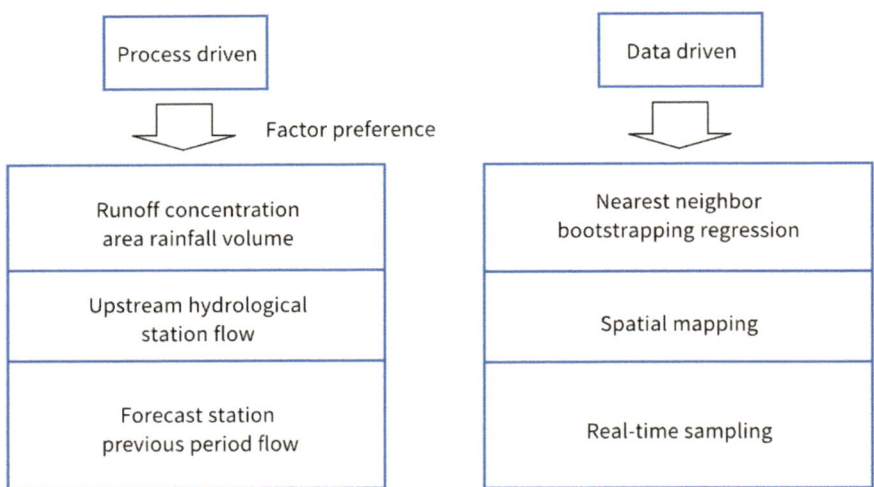

Fig. 4.4 Diagram of short-term runoff similarity forecasting

Short-term runoff similarity forecasting technology solves the problem of hydrological model applicability caused by the spatial and temporal variability of hydrological processes in complex watersheds, further raises the accuracy of inflow forecasting in the Dadu River watershed, and refines the hydrological forecasting system. The 3-day forecasting period forecast precision is shown in Fig. 4.5.

- Medium- and Long-term Hydrological Forecasting Based on "Quantity-Pattern" Similarity Theory

In terms of medium- and long-term hydrological forecasting, from the perspective of physical factors influencing future runoff in the Dadu watershed, factors related to earlier rainfall, earlier runoff, and atmospheric circulation are added in as factors influencing medium- and long-term runoff forecasting. Using stepwise regression theory, factors that act as indicators of the flow at different control cross sections

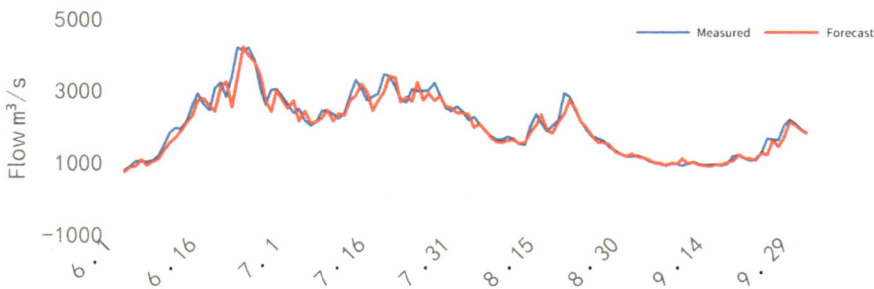

Fig. 4.5 The 3-day forecasting period forecast precision

of the watershed, with relatively high correlation, are selected from among earlier rainfall and runoff factors, and from among 130 atmospheric circulation factors. These factors are refined into 88 atmospheric circulation indexes, 26 sea surface temperature indexes, and 16 other indexes, and thus with this, the similarity factor index set is established.

Based on these similarity factor indexes, a medium- and long-term artificial neural network forecast model is established for the Dadu River watershed. "Quantity-pattern" similarity theory is added to objective function as an index assessment principle, so as to ascertain the earlier historical conditions closest to the forecast runoff, thus filtering out the historical runoff process closest to the forecast flow. Upon this, corrections to magnitude are carried out, and the forecast is realized.

The multifactor system for medium- and long-term forecasting for the Dadu River is shown in Fig. 4.6, and an example of medium- and long-term forecasting is shown in Fig. 4.7.

In summary, the integrated hydrometeorological forecasting system built by the Dadu River Company, with database-based comprehensive perception, watershed WRF model-based high-precision meteorological forecasting, deterministic hydrological model forecasting, probabilistic runoff forecasting, and data-driven similarity hydrological forecasting technology, allows realization of high-precision time, location, and quantity precipitation runoff forecasting within the scope of the watershed.

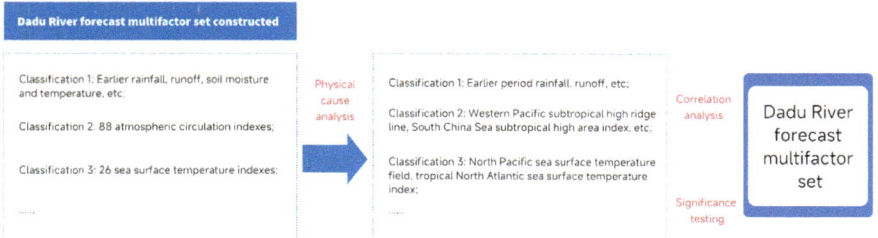

Fig. 4.6   The multifactor system for medium- and long-term forecasting for the Dadu River

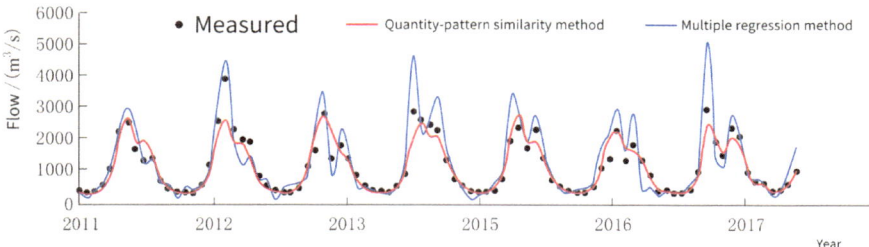

Fig. 4.7   Example of medium- and long-term forecasting

Although application of Dadu River hydrometeorological forecasting research has already achieved good results, there are still deficiencies in terms of integration of multi-source meteorological data, optimization of hydrometeorological forecasting models, and forecasting via multi-model advantage fusion. At present, improvements are being carried out around three areas. First, improving the precision and fineness of watershed status perception, implementing integration and assimilation based on the errors and advantages of multi-source data, and establishing a more refined and accurate basic materials library. Second, based on the existing meteorological forecasting model with a single, fixed parameterization solution, carry out clustering and separate optimization of the parametric solutions for typical rainstorms according to meteorological dynamic modalities, achieving a real-time forecast model with dynamically selected solutions according to meteorological dynamic modality representations. Third, starting from the perspective of atmospheric evolution mechanisms and runoff occurrence mechanisms, accomplish the deep integration of the two mechanisms, fully considering the impact of watershed hydropower station water storage and regulation, establishing a hydrometeorological forecast model for conditions under impact of high-intensity human activities.

4.2.2 Reservoir Joint Flood Control Smart "One-Button Dispatch" Technology

The successive commissioning of Dadu River mainstem and tributary reservoirs has appreciably improved runoff regulation capacity, but has also changed the hydrological and hydraulic properties of the watershed, increasing the complexity of the flood control, runoff concentration, and the hydraulic connections between the reservoirs of the cascade. This creates enormous challenges for the safe operation of the reservoirs during flood periods, bringing more meticulous demands to the work of reservoir dispatch. These new circumstances are difficult to meet through previous experience-based consultative decision-making mechanisms.

With that background, a cluster of optimal dispatch decision-making models was built for the reservoir group through system optimization, relying on water and rainfall monitoring and forecasting, equipment status perception, and automated control systems. New technologies were introduced, such as big-data and artificial intelligence, and a smart "one-button dispatch" technology was developed for reservoir group joint flood control (as shown in Fig. 4.8). This broke through the theoretical limitations and technical bottlenecks of the traditional reservoir dispatch decision-making model in terms of the utilization of forecast information, optimization of reservoir dispatch decision-making, and equipment control operation, and achieved the efficient coordination of multiple objectives such as flood control, benefit, and ecological protection in the watershed in different time scales.

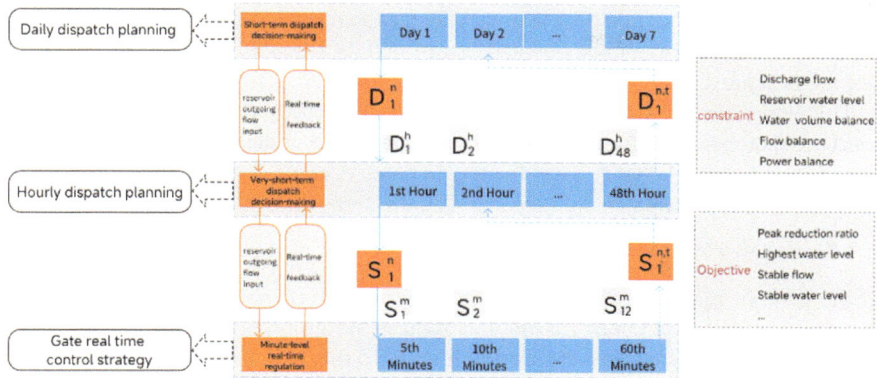

Fig. 4.8 Smart "one-button Dispatch" flood control technology

- Day-scale Short-term Dispatch Decision-Making

At the day-scale short-term dispatch level, a dispatch decision-making model was established with both flood control and benefit-raising modes. Based on the 7-day runoff forecast, the model determines whether the reservoir water level will exceed the water level limit during the forecast period, then selects a dispatch mode based on this. In flood control mode, the mathematical model is built with the objective of the lowest maximum reservoir water level during the forecast period. In benefit-raising mode, the mathematical model is constructed with the objective of the maximum power generation or the maximum reservoir water level during the forecast period. By updating the boundary conditions in real-time, the model smartly switches between flood control and benefit-raising modes, and automatically generates the dispatch decision solution for the cascade power plant group for the next seven days. This model translates the flood control demands of different levels, areas, and subjects into 112 water levels, flow rates, and their change rate constraints. By grading the constraints, the mathematical model can effectively identify the degree of importance of flood control objects upon encountering large floods, and generate corresponding dispatch strategies, improving the usefulness of flood regulation solutions.

The short-term dispatch decision model specifies the activation criteria for flood control dispatch. Based on this activation criteria, and through real-time smart mode switching, the most applicable dispatch mode can be activated at the most appropriate time, which maximally solves the contradiction between flood control and benefit raising, and maximizes the comprehensive effectiveness of the reservoir.

- Hour-scale Very-short-term Dispatch Projection

At the level of hour-scale very-short-term dispatch, a smart projection model for cascade reservoirs was established. Taking the end-of-day water level or power generation capacity of the short-term dispatch solution as a control condition, and combining in the hour-scale flood forecast and intra-day power generation plan, a real-time rolling projection of the cascade dispatch process, including reservoir water level, reservoir outgoing flow, power generation flow, flood discharge flow, and other factors, is achieved.

This very-short-term dispatch projection model performs simulation of intra-day hourly dispatch within the day-scale short-term dispatch decision framework. This provides an hourly dispatch basis for the dispatch execution layer, and achieves the organic nesting of dispatch solutions having different time-scales.

- Minute-level Real-time Dispatch Control

At the level of real-time dispatch, the first problem to be solved in order to realize minute-level dispatch control is how to reduce the minimum time scale for forecasting reservoir inflow from hours to minutes. To this end, a coupled hydrological-hydrodynamic model of each reservoir area was created for the Dadu River watershed, which established the basic prerequisites for the establishment of a minute-level optimal dispatch model. Based on the individual properties of various engineering entities such as overflow facilities and hydropower generator sets, and their mechanisms for cooperative operation, a real-time feedback and closed-loop correction strategy was introduced for water level deviation under high peak and frequency regulation conditions. A "gate + generator set" multi-unit cooperative multi-mode dispatch control model cluster was established with the goal of optimizing flood control effectiveness, overflow facility depreciation cost, and labor cost, forming a nested "short-term, very-short-term, and real-time" dispatch decision-making system for cascade reservoirs (hydropower plants), accomplishing effective and accurate execution of short-term, and very-short-term dispatch strategies. The minute-level real-time dispatch control model is shown in Fig. 4.9.

4.2.3   Smart Warning Technology for Reservoir Shoreline Side Slope Geological Hazards

Reservoir dispatch revolves not only around factors such as flood control safety and economic operation, but reservoir shoreline safety is also an issue that requires a high degree of attention since shoreline safety is an important prerequisite for safe operation of reservoirs. Therefore, it was necessary to establish a geological hazard survey and monitoring system covering the entirety of the administered range, so as to discover potential geological hazard points in a timely manner, conducting comprehensive monitoring of shoreline side slope geological hazards over a large area, and at a large scale. Effective technological means have been adopted in the Dadu River watershed to carry out focused control on the main places of concern discovered in the shore slope geological hazard survey, forming a multi-dimensional and multi-scale reservoir shoreline side slope monitoring model, from holistic to partial, and from large scale to fine. In temporal respects, not only is there instantaneous change monitoring of shore slope geological hazards, but there is also long-term time series trend

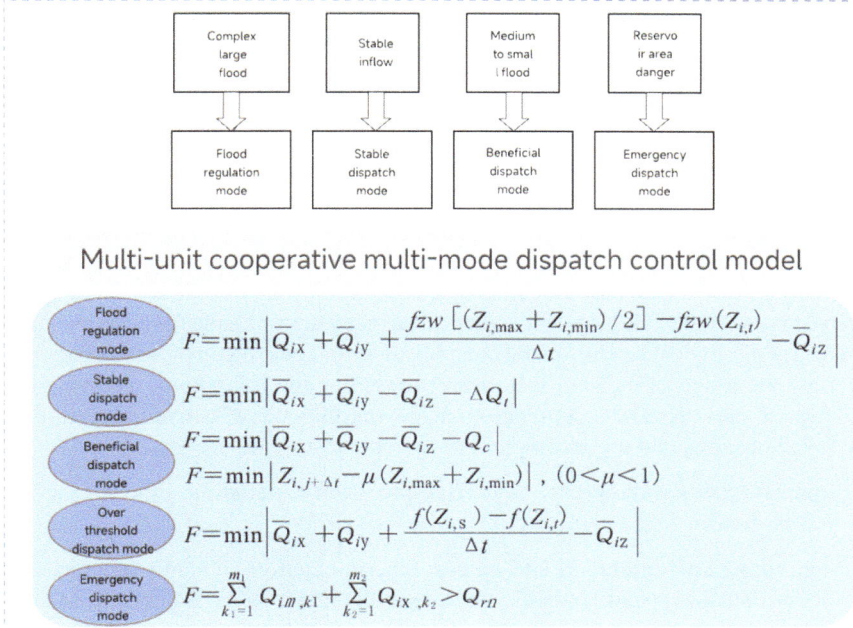

Multi-unit cooperative multi-mode dispatch control model

Flood regulation mode
$$F = \min \left| \overline{Q}_{ix} + \overline{Q}_{iy} + \frac{fzw \left[(Z_{i,\max} + Z_{i,\min}) / 2 \right] - fzw (Z_{i,t})}{\Delta t} - \overline{Q}_{iz} \right|$$

Stable dispatch mode
$$F = \min \left| \overline{Q}_{ix} + \overline{Q}_{iy} - \overline{Q}_{iz} - \Delta Q_t \right|$$

Beneficial dispatch mode
$$F = \min \left| \overline{Q}_{ix} + \overline{Q}_{iy} - \overline{Q}_{iz} - Q_c \right|$$
$$F = \min \left| Z_{i, j + \Delta t} - \mu (Z_{i,\max} + Z_{i,\min}) \right|, \ (0 < \mu < 1)$$

Over threshold dispatch mode
$$F = \min \left| \overline{Q}_{ix} + \overline{Q}_{iy} + \frac{f(Z_{i,s}) - f(Z_{i,t})}{\Delta t} - \overline{Q}_{iz} \right|$$

Emergency dispatch mode
$$F = \sum_{k_1 = 1}^{m_1} Q_{im,k1} + \sum_{k_2 = 1}^{m_2} Q_{ix,k_2} > Q_{rn}$$

Fig. 4.9 Minute-level real-time dispatch control model

change monitoring, as well as warning of geological hazards from external causes. In spatial respects, there are watershed-wide geological hazard monitoring systems, and also narrower monitoring systems looking at different zones or even working at the microscopic level. In dimensional respects, there is hazard monitoring based on singular geophysical properties, along with hazard monitoring based on multiple indicator systems, making it easier to grasp the evolution processes of the larger geological environment, and the patterns of occurrence and change of geological hazards.

(1) Early Identification of Geological Hazards

Based on "space-aerial-terrestrial" multi-source three-dimensional observation data analysis technology, a system for early identification of geological hazards was built. Highly suited for the Dadu River watershed, the system employs the overlapping triple assessments of initial survey, detailed investigation, and cross-verification. To address the high incidence of geological hazards on the shore slopes caused by the unremitting effects of rainfall and reservoir water level, as well as always-increasing human engineering activities, high frequency of extreme weather, and inadequacy of conventional inspection means, research was done on technological methods for conducting large-area geological hazard surveys based on satellite InSAR technology, and screening to come up with the focal deformation areas. Research has been carried out on applying airborne LiDAR-based technology for early identification of hazard object deformation markers and rapid capture of hazard precursor information, technological processes have been developed, such as hazard interpretation methods for airborne LiDAR images and UAV real-world 3D models, and detailed geological hazard surveys have been conducted of main suspect areas, achieving preliminary ascertainment of potential geological hazard sites. Careful verification of potential geological hazards is carried out through manual coordination to further confirm information on boundaries, scale, formation mechanisms, deformation markers, progressive stage, stability status, and threat range, to attain a refined differentiation of potential hazard sites. Upon this foundation, a multi-source three-dimensional observation data analysis technology is used in the Dadu River watershed to build a triple assessment identification system suitable for the early identification of geological hazards in the complex mountainous region of watershed, thus enriching the range of means for disaster prevention and mitigation, resolving the challenge of early hazard identification, achieving the goal of active prevention of geological hazards, and the preemption of major catastrophic events.

(2)   Landslide Whole-Process Rapid High-precision Simulation and After-effect Assessment

Through numerical simulation and indoor experimentation, dynamic mechanisms were identified for typical landslide hazards in the Dadu River watershed, and for the progressive dynamic processes of potential fracture planes. A rapid differentiation technology was developed covering the whole progression of sudden large-scale landslides. A cloud-based multi-source data mining technology was adopted to rapidly extract the structural properties and physical mechanics parameters of sloping rock and soil masses. A three-dimensional multi-scale refined model was established for very-large landslides. Utilizing a supercomputing platform, a high efficiency algorithm was created based on FEM/DEM and fluid dynamic (SPH/LBM) solid–liquid coupling. A strongly coupled computation method was invented for deformation, fracture, and movement during inclined slope hazard progressions, under multiple scales, and realizing high-performance, refined, and rapid computational analysis for very-large landslide masses.

In terms of quantitative risk assessment, a technology based on machine learning was created for the rapid retrieval of physical mechanics computational parameters for very-large landslide masses, able to quickly call up parameters related to analogous hazard cases from the cloud. For very large very-large landslides masses, a rapid and refined simulation based on solid–liquid coupling can be carried out, covering the whole process of "deformation $\rightarrow$ progressive damage $\rightarrow$ catastrophic fracture $\rightarrow$ rapid displacement $\rightarrow$ accumulation," achieving an efficient and rapid quantitative risk assessment. Directed at the specifics of the multiphysical processes of the hazard chain, a physical model of geological hazard chain dynamics progression in real spatial scale was established, and a numerical simulation platform for hazard chain dynamics progression was developed (with independent intellectual property rights), maximizing the quantitative assessment of risk.

(3)   Real-time Collection of Multi-source Heterogeneous Monitoring Data

The types of data from shore slope geological hazard monitoring are diverse, and there are certain differences in regard to data acquisition, transmission, and storage, so the integration of the various forms of monitoring data becomes quite important. For multiple source heterogeneous monitoring data, a set of real-time integration technologies was implemented in the Dadu River watershed to integrate monitoring data of various types and from different instruments into the real-time monitoring center database to achieve rapid integration of multi-source heterogeneous monitoring data. In general, the acquired monitoring data (raw data) cannot be directly used for early warning computations, and pre-processing of monitoring data is still required. Carrying out research on the processing methods for missing data, anomalous data, and noisy data, and conducting adaptability analysis on each data processing method, has led to completely automated programmatic pre-processing of monitoring data. The application of real-time integration technology for multi-source heterogeneous geological hazard monitoring data achieves automated real-time integration of multi-source heterogeneous geological hazard monitoring data into the database, along with processing, filtering, and analysis of anomalous monitoring data.

At the same time, a multi-source heterogeneous data integration platform has been developed to utilize data services to integrate monitoring data from different manufacturers and equipment into a unified monitoring database, thus achieving the integration of multi-source monitoring data.

• Reservoir Shoreline Side Shore Safety Early Warning for Rainfall Causes

Deformation failure in reservoir shoreline landslides is a complex geological process, occurring under the cumulative effects of geological environment factors (such as geological structure, terrain and geomorphology, formation lithology, etc.), and external factors (such as atmospheric rainfall, reservoir water level fluctuation, human engineering activities, earthquake, etc.). Looking at Dadu River historical reservoir shoreline monitoring data, the geological environment in almost invariably comparatively stable, and external factors are the primary triggers for reservoir shoreline geological hazards.

In order to further improve command of reservoir shoreline safety, cross-specialty integration of "rainfall forecast + geological hazard warning" was implemented, based on multi-source meteorological forecasting and a network for reservoir shoreline side slope displacement monitoring. Combining in landslide displacement monitoring data, and from the perspective of analyzing the progression pattern of landslide displacement, landslide displacement is broken down into the two categories of trend displacement, which is controlled by its own foundational geological conditions, and periodic displacement, which is impacted by external triggering factors such as rainfall. A forecast model is constructed based on displacement category mined from data, and total displacement prediction is obtained by superimposing results for both categories.

This technology enables reservoir shoreline safety warning work to achieve full coverage spatially, starting with individual points and extending to the whole, going from measuring rain that has fallen to forecasting rain that is coming for the forecast period, providing more comprehensive and more accurate rainfall geological hazard warning services for watershed reservoirs, local towns, road traffic management, and other departments.

- Building a Platform for Smart Forecast and Early Warning of Geological Hazards

The general approach for exploring intelligent forecasting and warning of geological hazards in the Dadu River watershed is to establish a database for watershed geological hazard monitoring data, and carry out integration of the specialized monitoring data, so as to provide supporting data for subsequent warning analysis. Therefore, a fully built platform is a platform established upon geological hazard survey results, combined with geological analysis and monitoring and early warning, and jointly assembled by "industry, academia, research, users, and management" and other entities concerned with geological hazard prevention and control.

With this in mind, the Dadu River watershed geological hazard forecast and early warning platform was developed based on research into a model covering both general landslide warnings and general mudslide warnings. The platform's main functions include: 3D comprehensive information display, comprehensive information control, monitoring data control, automated geological hazard identification, geological hazard risk ranking, hazard chain whole-process simulation, smart early warning control, and assisted decision-making.

The 3D comprehensive information display module mainly provides graphic services and attribute services related to major geological hazards. Graphic services include display of geological hazard-related graphics, query of attribute information, positioning, and other operational functions. Attribute services mainly provide query and editing functions for geological non-spatial hazard data through an external attribute database.

The comprehensive information control module primarily functions to integrate spatial attribute information of major geological hazards across watershed hydropower stations, with efficient input, editing, and management. Essential map functions and operations, layer management, etc., are all supported, with the ability to superimpose on maps such items as hazard points, monitoring points, high-definition remote sensing images, geological maps, etc. This module also provides user role category management.

The monitoring data control module handles daily data management, and is mainly focused on multi-source heterogeneous monitoring data, achieving automated real-time integration into the database of multi-source heterogeneous geological hazard monitoring data, while filtering and analyzing anomalous monitoring data, and also carries out extended support to other databases.

The automated geological hazards identification module utilizes InSAR, optical remote sensing, UAV digital photogrammetry, and airborne LiDAR technologies to obtain a range of spatial data for watershed hydropower stations. With this foundation, a high-precision, high-speed, and high-depth neural network statistical model based on deep learning was built to achieve early automated identification of geological hazards in the watershed. Through the collation and analysis of multi-source data and model calculation, the scope of on-the-ground verification is quickly determined, the purpose, principles, and methods of detailed on-site investigation of suspected potential hazard points are determined, and the "triple assessment" technological system of initial survey, detailed investigation, and cross-verification for early identification of geological hazards in the watershed is established.

The geological hazard risk ranking module determines a quantitative assessment model for landslide geological hazard risk in the watershed based on previously researched key indicator factors for the assessment of susceptibility and risk for the landslide geological hazard points of watershed hydropower stations. Landslide geological hazard risk ranking impact factors are extracted based on previous landslide geological hazard occurrences, and the quantitative assessment model of the landside geological hazard risk for the hydropower stations in the watershed is integrated with the risk ranking model to build a dynamic risk ranking system for the geological hazards in the watershed, and on-site risk verification is carried out according to the results of this ranking system.

The hazard chain whole-process simulation module employs 3D digital earth animated projection and reconstruction of the geological hazard process to directly demonstrate the geological hazard triggering conditions, event process, and impact scope. This is accomplished by combining existing research results and field investigation regarding geological hazard causes, regional geological context, geological hazard monitoring and early warning, and through a self-developed 3D landslide disaster whole-process high-precision rapid simulation model. At the same time, based on detailed field investigation, the hazard chain whole-process simulation module can achieve projection of hazard processes for concealed potential geological hazard sites, and the reconstruction of previous geological hazard events.

Based on the full utilization and integration of existing achievements in geological hazard prevention and control, and by making comprehensive use of big-data, GIS, geological hazard space-aerial-terrestrial detection, 3D visualization, cloud computing and other technologies, a geological hazard prevention and control platform suited to the characteristics of the Dadu River watershed has been built. A big-data resource pool of geological hazard prevention and control information gathered from multiple channels was built, and a full-coverage technical support platform and methodological system was established for early identification, risk ranking, monitoring and early warning, and emergency command of geological hazards. Thus, the informationized, intelligent, and standardized management of the whole process has been achieved, from data aggregation, data management, risk assessment, monitoring and early warning, command, and dispatch, to comprehensive prevention and control.

4.3 Case Application

4.3.1 Successful Response to a 100-Year Major Flood on the Upper Reaches of the Dadu River

On June 13, 2017, 48 h in advance, the Dadu River watershed high-precision hydrometeorology forecasting system accurately forecast a mega-flood on the upper reaches of Dadu River. The system issued timely warning alerts and decisively pre-discharged and lowered reservoirs, and through the newly built Houziyan Reservoir, ultimately held back the flood waters and successfully met the challenge of the "Danba County 6.15" 100-year flood in the upper reaches.

- Accurate Forecasting of Rainfall and Water Inflow Trends 48 h in Advance

On June 12, 2017, numerical forecast results of the Dadu River hydrometeorological forecast system showed that the upper reaches were expected to face a round of heavy rainfall from June 13 to 15, and the cumulative surface rainfall above Danba County was expected to exceed 30 mm. Through the Xin'an River and API (Antecedent Precipitation Index) multi-model forecasting, and combined with manual verification, consultative forecasting for the hourly and daily cross-sectional flow for the Danba section from June 13–16 was carried out, with the cross-sectional flow forecast to reach a peak of 5,000 cubic meters per second on the 15th, and the rainfall area expected to shift downward and rainfall intensity decrease, with later water inflows predicted to decline.

- Relentless Heavy Rainfall, a Rare 100-year Flood as Expected

From the start on June 13, the forecast and actual conditions were very similar. On the 13th, 14th, and 15th, in the upper reaches the daily rainfall was 18.8, 16.6, 3.0 mm, respectively, with a three-day cumulative rainfall of 38.5 mm. Forecast accuracy was 95.2%, and strong rainfall was mainly concentrated in the area above Danba, whereas rainfall in the area below Danba gradually weakened. Comparisons of forecast and actual rainfall can be found in Fig. 4.10.

In this instance, rain fell widely across the watershed, and the time frame of initial rise showed uniformity across stations; flooding was primary riverine flooding (Table 4.2).

- Flood Detention and Peak Shaving to Ensure Safety of Two Cities"

In the middle reaches of the Dadu River watershed, there is the historical city of Luding County, renown in particular because of the Luding Bridge, along with the Shimian County national key ecological function zone. Without the storage and regulation action of the reservoir, impounding water and shaving the flood peak, and if the unimpeded natural floods had been allowed to rush and rampage down through the watershed, then the important and densely populated towns of Luding and Shimian along the middle reaches would have suffered unimaginable loss of

Fig. 4.10 Top: Actual
rainfall across three zones of
the watershed. Bottom:
Comparison between
forecast and measured
rainfall in the watershed

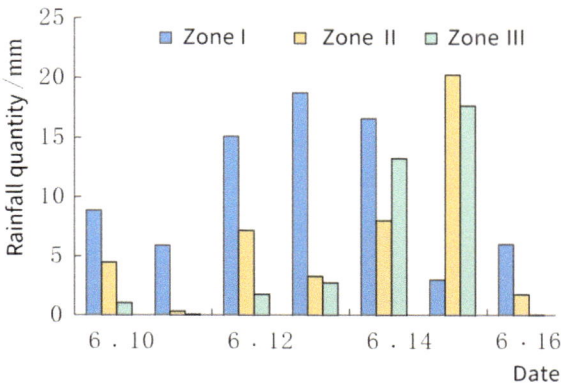

(a) Actual rainfall across three zones of the watershed.

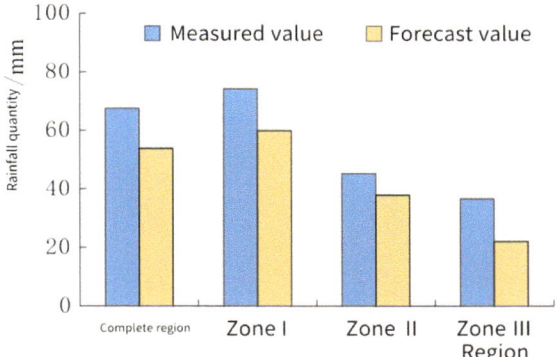

(b) Comparison between forecast and measured rainfall
in the watershed

Table 4.2 Flood conditions at key measurement stations

Sequence	Measurement station	Time of initial rise	Time of peak	Peak flow (m³/s)	Amount of increase (m)
1	Ribu hydrological station	June 11, 05:00	June 14, 22:00	1560	1248
2	Zumuzu hydrological station	June 11, 05:00	June 15, 02:00	2222	1248
3	Dajin hydrological station	June 11, 04:00	June 15, 20:00	3795	2963
4	Danba hydrological station	June 11, 13:00	June 15, 15:00	4990	3505

life and property, including even the inundation of the county seats. Local economic development would have been greatly affected. The relative locations of Houziyan hydropower station, Luding County, Dagangshan hydropower station, and Shimian County are shown in Fig. 4.11.

In the face of this 100-year flood, the Dadu River Company actively lived up to its social responsibilities, providing forecasts and warnings in advance, establishing a dispatch contact mechanism with local governments and flood control authorities at all levels, taking full advantage of the storage and regulation, impoundment, and detention actions of the newly commissioned Houziyan Power Station reservoir storage and flood control. Cumulative impounded flood volume was 211 million cubic meters. This deeply reduced the flood peak, effecting a significant reduction compared to the unimpeded natural flow. The highest rate of peak shaving was 24.9%, lowering the peak flow level at the Luding and Shimian county seats by 1.5

Fig. 4.11 Relative locations of Houziyan hydropower station, Luding County, Dagangshan hydropower station, and Shimian County

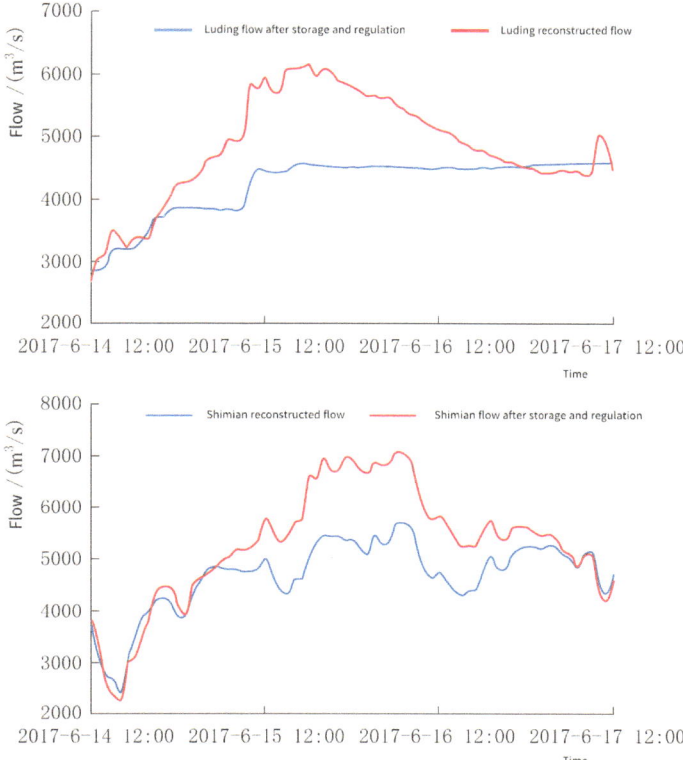

Fig. 4.12 Comparison of reservoir inflow volume at Luding and Shimian after storage and regulation at Houziyan

and 1.4 m, respectively. A 100-year flood was converted into a 15-year flood. After storage and regulation, the lives and properties of people along both sides of the river were successfully protected, and the flood waters smoothly passed their way through Luding and Shimian. The storage and regulation process is shown in Fig. 4.12, and flood prevention results are shown in Table 4.3.

4.3.2 Successful Response to a 100-Plus-Year Flood on the Middle and Lower Reaches

In 2020, the Changjiang River watershed experienced heavy rainfall over a wide area, with intense rainstorms. Affected by repeated rounds of heavy rain upstream, flood waters on the Changjiang manifest high flood peaks, high volumes, forceful rise, and great destructiveness. In terms of flood water regulation in this case, through accurate

Table 4.3 Flood peak shaving results from storage and regulation by Houziyan power station during the 2017 "6.15" flood

Station	Natural peak volume (m³/s)	Corresponding natural water level (m)	Peak volume after storage and regulation (m³/s)	Corresponding peak level after storage and regulation (m)	Peak reduction ratio	Level reduction at downstream stations (m)
Luding	6140	1313.4	4610	1311.9	24.9%	1.5
Shimian	7280	855.1	5680	853.7	22.0%	1.4

plandslide was located onrediction, scientific dispatch, and repeatedly taking advantage of the Pubugou hydropower station—the Dadu River's dispatch "general on/ off switch"—to carry out pre-discharge, reservoir lowering, flood water impoundment, and peak staggering, to successfully respond to repeated large flooding, greatly reducing flood water pressures on the Dadu River as well as along the middle and lower reaches of the Changjiang River. Especially in this 100-plus-year large flood, the Dadu River Company made outstanding contributions in disaster prevention and mitigation.

(1) Accurate Forecast of Rainfall Trends 96 Hours in Advance

On August 12th, seven days before the flood peak, through the application of core technologies such as high-precision coupled forecast for water and rainfall, refined simulated projection, and multi-scale flood regulation decision-making, it was predicted in advance that powerful rainstorms would continue from the 15th to 18th (as shown in Fig. 4.13), and it was expected that the stretch from Ebian to the Gongzui and Tongjiezi power stations might suffer an exceptional flood, with a peak approaching 10,000 m³/s. Rainfall and runoff forecast results were immediately released to all units in the watershed. Based on the simulated dispatch solution projected by the system, there was a full deployment for flood prevention work, the whole dispatch line mobilized to enter into large flood prevention "war time" footing, standing ready to deal with a 100-plus-year flood.

(2) Flood Water Regulation Process and Results During the "8.18" Mega-flood

August 16, three days before the flood peak, Pubugou reservoir was dispatched to pre-discharge to a level two meters below the flood-limited water level (FLWL), striving to free up reservoir capacity for later impoundment of flood water.

At 18:00 on August 17, the Pubugou Dam spillway tunnels were closed completely. Only 2300 m³/s was being discharged, to maintain power generation. Starting on the morning of the 18th, reservoir inflow continued to rise to more than 6000 m³/s, with a maximum peak of 6991 m³/s. The maximum volume of peak reduction reached 4691 m³/s.

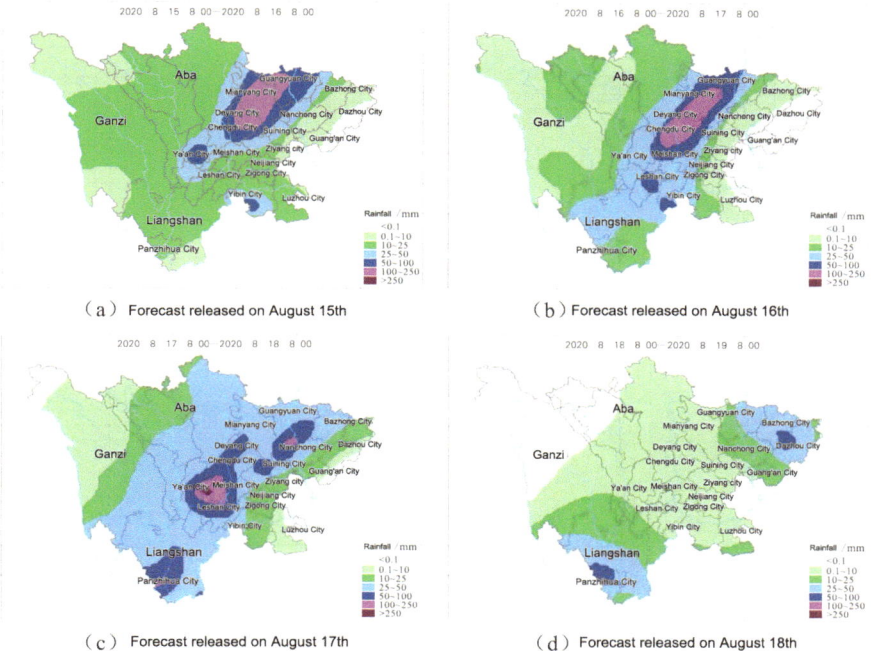

Fig. 4.13 Rainfall predictions for August 15–18 for areas below Houziyan

After regulation and storage at Pubugou, at 06:00 on August 18, flood natural peak flow downstream at the Gongzui power station had been reduced from 12,600 to 6740 m³/s. At 09:00 on the same day, natural peak flow downstream at the Tongjiezi power station had fallen from 13,300 to 6560 m³/s. Cross-sectional flooding at locations like Ebian, Gongzui, and Tongjiezi had transformed from an exceptionally large 100-plus-year periodic flood, into a common flood. Flood peak conditions at main cross-sections are shown in Table 4.4. Relative locations of power stations and towns in the middle and lower reaches of the watershed are shown in Fig. 4.14.

Without accurate advance anticipation, and without timely and decisive decision-making and scientific dispatch, locations on the lower reaches of the Dadu such as the area around Leshan City, the Jinkouhe District, Ebian County, and the Shawan District would have been flooded by nearly 3 m of water. Nearly 50,000 people would have suffered the ravages of flooding, and the inundations of the city proper area of Leshan City, as well as Changjiang River areas of Sichuan and Chongqing, would have been even more calamitous. For this reason, the Ministry of Water Resources of the People's Republic of China sent a letter commendation to the Dadu River Company for its significant contributions to flood prevention and mitigation in Sichuan Province and throughout the Changjiang River watershed.

Table 4.4 Flood peak conditions at main cross-sections during the "8.18" flood period

Cross section	Peak volume (m³/s)	Flood peak arrival time	Periodicity	Flood peak after storage and regulation (m³/s)	Natural peak volume (m³/s)	Natural periodicity
Pubugou	6990	Aug. 18, 03:00	20-years			
Gongzui	7810	Aug. 18, 06:00	7-years	6740	12,600	100-plus-years
Tongjiezi	7430	Aug. 18, 09:00	5-years	6560	13,300	100-plus-years
Yanrun station	1430	Aug. 18, 09:40				
Hongqi station	1780	Aug. 18, 05:45				

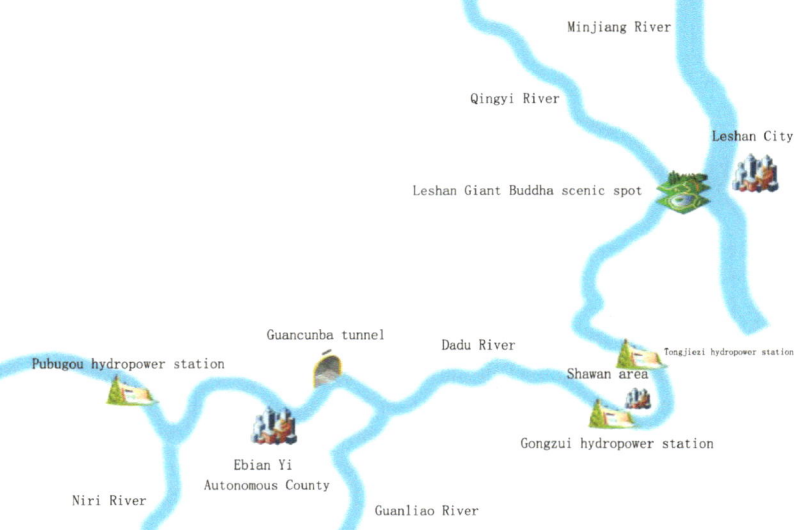

Fig. 4.14 Location of power stations and towns in the middle and lower reaches of the Dadu River watershed

A comparison between reservoir inflow and the reconstructed flow at the Gongzui and Tongjiezi stations is shown in Fig. 4.15.

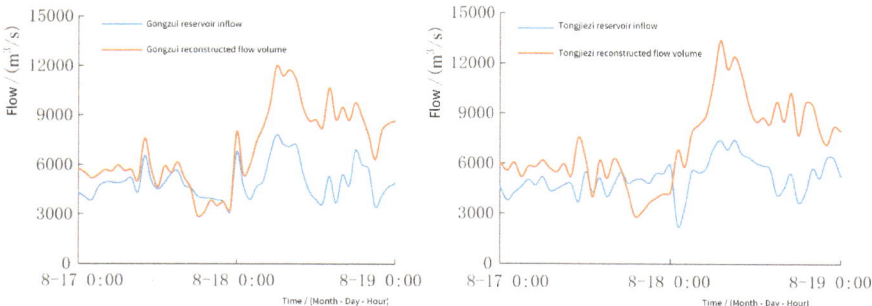

Fig. 4.15 Measured reservoir inflow volume and reconstructed flow volume for Gongzui and Tongjiezi stations during the "8.18" flood process

4.3.3 Accurate Warning of Major Collapse of the Kaiding Landslide in the Houziyan Reservoir Area

- Basic Conditions

The Kaiding landslide was located on Provincial Road S217 in Danba County, Ganzi Prefecture, Sichuan Province, in the reservoir area of the Houziyan hydropower station. The total volume of the slide mass was about 4.5 million cubic meters. In January 2018, geological hazard safety risk perception data showed an uncharacteristic increase in the rate of deformation of Houziyan hydropower station's rerouted Provincial Road S217. The geological hazard smart control platform then issued a risk early warning, and implemented graded safety risk control of this side slope. At the same time, supplementary multi-source information collection devices such as smart sensors, mounted mini-sensors, 3D laser scanning, and smart inspection UAVs were added to perceive the operational status of the slope in real-time, and big-data acquired through perception was uploaded to the geological hazard smart control platform in real-time.

- Risk Early Warning and Results

Relying on the geological hazard smart control platform and applying the geological hazard risk assessment model cluster, and combining the results of numerical analysis calculation, and integrating multi-source data fusion analysis, along with overall research and judgment, led to a conclusion: The deformation rate of the 3.5 million square meter area Kaiding landslide mass would further accelerate, and a large-scale collapse was estimated to occur when the deformation rate exceeded about 50 mm/day. At the same time, the deformation rate of the landslide mass was positively correlated with the rate of water level drop, thus the rate of water level drop required strict control.

On February 9, 2018, the displacement rate of the Kaiding landslide mass reached 50 mm/day, and the system issued a warning message. Road traffic controls were implemented, and personnel evacuated. A large-scale landslide occurred on the slope on February 13, with the complete mass sinking from 3 to 8 m. The Kaiding landslide and its slope displacement rate are shown in Figs. 4.16 and 4.17. Due to the timeliness of the warning, potential casualties and property damage caused by the landslide were avoided.

Additionally, employing the big-data platform, analysis based on multi-source long series data showed that although the deformation rate of the Kaiding landslide mass had decreased, there was still a risk of continued collapse, and this circumstance did not allow immediate clean-up and rectification work. At the same time, in order to meet the transportation needs of the local people to pass through the slide zone, the Dadu River Company developed real-time analysis of geological hazard monitoring big-data and opened up a temporary transport passage for use as long as conditions were safe.

Fig. 4.16   Image of real-time monitoring for the Kaiding landslide

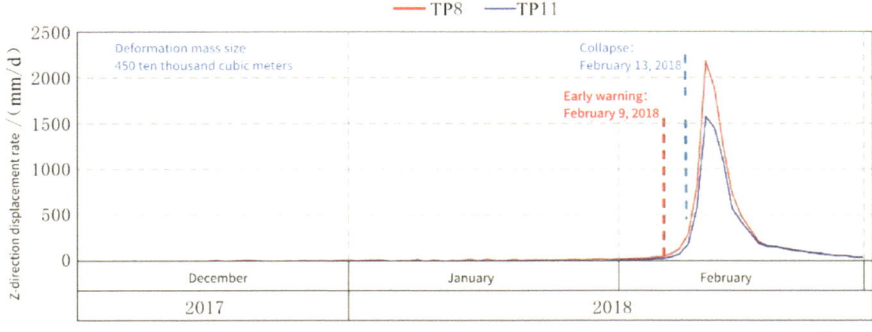

Fig. 4.17   Side slope displacement rate of the Kaiding landslide mass

4.3.4 Accurate Warning of Collapse of the Huangcaoping Deformation Mass in the Dagangshan Reservoir Area

- Basic Conditions

The Huangcaoping deformation mass, with a volume of about 1.5 million cubic meters, is located along the shoreline of the Dagangshan hydropower station reservoir, as shown in Fig. 4.18. In June 2018, through big-data sensing and early warning analysis, the trend of slope deformation was judged to indicate clear risk of large-scale collapse. Nine days after the release of early warning information, a large-scale collapse occurred at this deformation site. Due to the appropriate early warning and handling measures, casualties and property losses caused by the landslide were avoided.

- Risk Discovery

From July to November 2018, the deformation of the slope mass increased, and partial collapse occurred. Reservoir and dam safety risk perception data showed that the deformation rate of the monitored area of the deformation mass was increasing abnormally. The safety risk smart control center then issued a risk early warning, implemented graded safety risk control of the slope, and turned on four smart perceptive devices to continuously monitor the deformation trend of the deformation mass 24 h a day. Supplemental mini-sensors, smart inspection UAVs and other multi-source information acquisition equipment was added to perceive operational status of the slope in real-time, and upload the data obtained through perception to the safety risk smart control center in real-time.

In reliance on the reservoir and dam risk early warning analysis platform, and utilizing support libraries driven by three major smart inference systems, employing a

Fig. 4.18 The Huangcaoping deformation mass monitoring screen

dynamic safety assessment model based on information entropy, big-data cooperative processing and overall analysis of integrated multi-source data by the safety risk smart control center showed that the deformation rate would further accelerate, with large-scale collapse considered likely when the deformation rate exceeded 50 mm/day.

- Response Measures

Based on early warning results, decisive response measures were taken:

(1) Prompt formulation and activation of emergency contingency plans for geological hazards.
(2) Initiation of controlled- and restricted-access measures to ensure the safety of passing vehicles and pedestrians.
(3) Encrypted monitoring and real-time analysis of the slope based on a big-data evolutionary inference analysis model, determining chronological and spatial change trends of monitoring data, analyzing for unusual changes in the landslide mass, and warning well in advance of slope collapse risk.
(4) Prompt research on a management solution for the deformation mass, and the formulation of temporary measures to preserve transportation.

- Getting Results

On June 25, 2018, the rate of deformation at Huangcaoping reached 447 mm/day and a partial collapse occurred. On July 2, collapse occurred over a large area, with a collapse volume of about 50,000 cubic meters. About 100 m of roadbed collapsed or suffered damage. Forecast and early warning for this collapse applied big-perception, big transmission, big-storage, big-computing, and big-analysis technologies. Early warning information was issued 9 days in advance, on June 16, and prompt implementation of measures to control access and evacuate personnel averted landslide casualties and property damage.

After the large collapse, using the big-data platform, analysis based on multi-source long series data showed that although the deformation rate had decreased, there was still a risk of continued creep. Supplemental micro-sensor systems and four additional smart perception devices were set up. At the same time, in order to satisfy the transportation and daily life needs of the local people, the Dadu River Company established real-time analysis of geological hazard monitoring big-data, and opened up a temporary substitute pedestrian passage, with access controlled dynamically based on safety conditions.

4.3.5 Scientific Management of the Zhengjiaping Deformation at the Dagangshan Power Station

(1) Basic Conditions

The Zhengjiaping deformation is located along Provincial Road S217 in the reservoir area of the Dagangshan hydropower station. As shown in Fig. 4.19, the total volume of the deformation mass is more than 3.2 million cubic meters. Examination in February 2016 revealed cracks within the Zhengjiaping deformation area. The Dadu River Company worked jointly with a design unit and together set up 14 smart perception devices to continually monitor the progression trend of the deformation 24 h a day, while at the same time uploading in real-time the big-data obtained through perception to the geological hazard smart control platform. On April 30, 2016, the eve of the busy May Day national holiday, a large-scale collapse of the slope was accurately forecast 4 h in advance. Prompt release of warning information and decisive access-control measures successfully prevented major casualties and property losses on the lifeline Sichuan-Tibet Highway during the high traffic conditions of the holiday.

(2) Science Determines the Solution

After the initial partial collapse, big-data perception and early warning analysis indicated that the overall stability of the deformation mass was poor, and if deformation continued to progress at a later period, then soil collapse or landslide could occur, endangering the safety of transportation on Provincial Road S217, activities on the water, and operation of the power station, etc. Because of this, prompt research was carried out on a management solution for the deformation mass, and a long-term

Fig. 4.19 Screen grab of the Zhengjiaping deformation warning monitoring screen

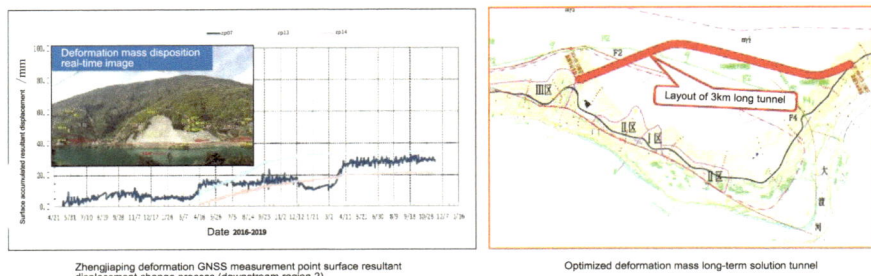

Zhengjiaping deformation GNSS measurement point surface resultant
displacement change process (downstream region 2)

Optimized deformation mass long-term solution tunnel

Fig. 4.20 The optimized transportation solution for the Zhengjiaping deformation at Dagangshan

management program was formulated. Through multiple discussions and reviews organized by the Dadu River Company, management principles were confirmed as (1) priority provisional implementation of a slope remediation solution, and (2) as conditions warranted, construction of a traffic tunnel in a separate location as a long-term solution.

(3)   Rational Optimal Design

Under the effective guidance of big-data perception and early warning analysis technology, in May 2017 work on provisional remediation of the deformed slope was safely and successfully completed, ensuring the safety of vehicles and people passing through.

Throughout two years of close monitoring and analysis during the flood seasons of 2017 and 2018, the deformation body maintained greater stability overall. Through Dadu River Company-organized expert review and deliberation, it was decided to cancel the proposed 3-km-long traffic tunnel solution, saving about 180 million yuan in costs. The optimized transportation solution is shown in Fig. 4.20.

Chapter 5
Intelligent Watershed Hydropower Dispatch

5.1 Approach and Goals

By constructing a large-perception system for water conditions, weather, equipment, and market power supply and demand, creating a series of analysis and prediction models for factors influencing the power market, such as hydrometeorological conditions, equipment health status, and marginal clearing price, and developing an optimization model applicable to cascade hydropower generation bid offering solutions, and developing a smart "one-button" dispatch model for cascade power stations, then the goal of more comprehensive perception of the market environment, more scientific market trade decision-making, and smarter regulation of cascade power generation in the Dadu River watershed can be achieved.

The approach to intelligent power dispatch on the Dadu River is shown in Fig. 5.1 and is composed of the primary constituent elements, below.

Market Environment Perception: Perception of the electricity market environment is an important part of market trading. Perception of the Dadu River watershed power market environment mainly includes supply, demand, and trade information perception. Among these, supply perception information includes information on historical power generation, inter-regional power purchases, water inflow forecasts, etc. Demand perception information includes historical power consumption, inter-regional outward transmission, bulk commodity trading, weather forecasts, etc. Trade information perception includes historical market power trading volume and price, market demand forecast curve, and other information released by the market. At the same time, utilization of big-data technology to analyze the perceived information, research and conclusions on market supply and demand conditions, and forecast of the marginal clearing price, provide important substantiation for power market trade decision-making.

Optimization of Trade Solution: The Dadu River Company has established an optimization calculation model for maximized power generation benefit, obtaining optimal and good trading solutions through optimization calculations. This happens

Y. Tu, *Management of Hydropower Enterprises*, Water Resources Development and Management, https://doi.org/10.1007/978-981-97-5584-4_5

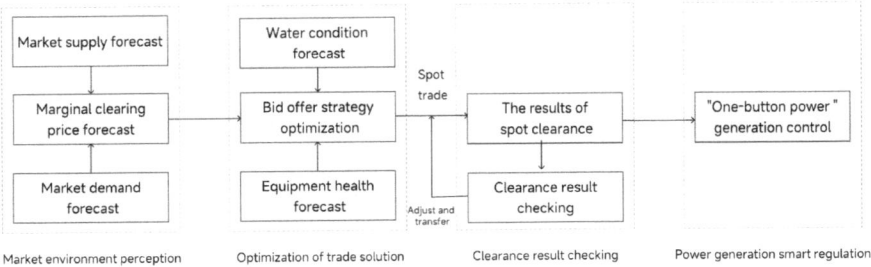

Fig. 5.1 Approach to intelligent power dispatch on the Dadu River

under the prerequisites of safe and stable operation of the dam and reservoir shoreline side slopes, with boundary conditions being the results of hydropower plant water forecasting (see Chap. 4), the results of equipment health and availability analysis (see Chap. 6), the results of marginal clearing price analysis and forecast, and the reservoir water level control target. In spot trading, multiple volume-price combination ranges can be declared for each time period, and power generating enterprises can develop their own particular offering strategies and corresponding trading solutions according to market circumstances and risk preferences.

Clearance Result Checking: After market trading, the clearance may result in a misalliance between cascade stations, and electrical generation load may fall into the rough zones for power station generator sets. For this reason, the Dadu River Company carries out simulation calculations on the clearance result to investigate its safety and economics, to answer questions such as: Will operation in the rough zone continue for a long period? Will the stability of the reservoir shoreline side slope be impacted? Will flood control and water supply safety be influenced? If serious problems are discovered, then prompt appeal can be made. It is also necessary to scrutinize the economics of the clearance results of upstream and downstream hydropower stations: Will there be water spillage losses? Will the reservoirs operate at low water level for a long period? Adjustments can then be made through medium- and long-term electricity transfer transactions or real-time spot trade.

Power Generation Smart Regulation: A tight hydraulic interconnection exists between individual cascade hydropower stations. Scientific arrangement of the power generation operation mode for cascade power stations is not only able to reduce water spillage and reduce the water consumption rate, but can also enhance the supporting effect of power stations for the safety and stability of the power grid. The Dadu River Company applies real-time optimized smart regulation technology for cascade power station loading and promptly optimizes power generation control for cascade power stations according to the results of spot clearance or according to power grid peak regulation and frequency regulation instructions. This reduces labor intensity for dispatch personnel and decreases human interventions, making cascade power generation safer, more economical, and more scientific.

5.2 Key Technology

5.2.1 Real-Time Smart One-Button Load Dispatch Technology for Cascade Power Stations

In the Sichuan power gird, Dadu River watershed power stations carry the primary responsibility for peak and frequency regulation. This real-time regulation is not only associated closely to the safe and stable operation of the power system, but is also intimately associated with the water conditions at cascade power stations. Yet, with traditional automatic generation control (AGC) technology, balancing out and economically controlling the load between individual cascade hydropower stations is difficult, making it extremely easy for water levels to rise and fall in large swings, or for spillage or reservoir emptying, seriously affecting the safe and economical operation of cascade power stations. Since real-time load regulation has numerous constraints and high real-time demands, this makes the search for a solution even more complicated.

Given all of this, the Dadu River Company considered the influence of multiple aspects, such as related to the power grid, reservoirs, generator sets, and economical operation, and built a set of real-time smart load regulation models suitable for cascade power stations as the primary force behind peak and frequency regulation, and developed a real-time smart load distribution system for watershed cascade power plants, achieving real-time smart regulation of cascade power plants under plant-grid coordination mode.

• Model Design

Under China's currently implemented power system automatic generation control (AGC) architecture, real-time load distribution of cascade hydropower stations should be accomplished at the intermediate layer between the power grid AGC and the power station AGC, and while first under a prerequisite of ensuring the safety of the power system, achieving economical cascade hydropower station dispatch control. The overall approach is that, based on system load demand or frequency change conditions, in real-time the power grid dispatch center gives overall power generation load commands to the cascade real-time smart load distribution system. Then in real-time, while also monitoring the operational status of each power station's AGC, the cascade real-time smart load distribution system accepts the overall power grid power generation load command and successfully divides up and allocates this overall load between the power plants, and sends the result of this load allocation to the AGC of each power station. The AGC of the individual power station is then responsible for the load distribution between generator sets within that plant, and returning feedback on execution results (as shown in Fig. 5.2).

• Modeling Strategy

In the real-time dispatch for cascade hydropower stations, the first priority is safety, and the second is economics. Safety is mainly reflected in two areas: First, ensuring

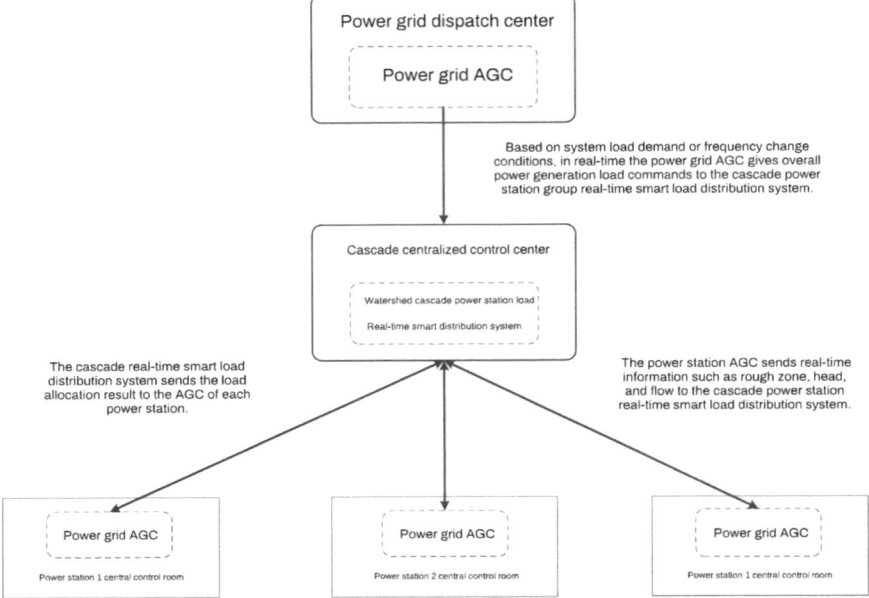

Fig. 5.2 Relationships between cascade centralized control center load distribution dispatch commands

both power grid frequency stability and prompt response to peak and frequency regulation demands of the power system; Second, attending to safe operating constraints such as power station operating water levels or minimum water discharge flow rates. Economy is reflected in areas such as whether there are water spillage losses in the cascade, whether the watershed water consumption rate is optimal, and how many times the generator sets are adjusted (which impacts set wear).

Based on the safety and economic requirements above, cascade real-time smart load regulation technology divides real-time load regulation into two major categories: command mode and non-command mode. In command mode, when the power grid AGC issues a load adjustment command, the load distribution system automatically pairs up a fast adjustment model, based on the principle of fastest time to overall load adjustment, such that the necessary time for all power stations to together complete the power grid load adjustment command is the shortest possible, satisfying system peak and frequency regulation needs. After the overall load adjustment is in place, and while ensuring that the total load stays relatively steady, the load is transferred between stations through simultaneous back and forth adjustment, the load of each station distributed in a way conducive to the economic operation of hydropower stations, ensuring the efficient utilization of water potential.

In the non-command mode, in regard to the discrepancy between the model's calculated results and actual dispatch results, the operational water level range is divided into three subdivisions: high water level operating zone, normal operating zone, and dead water level operating zone. When the water level enters either the high water level or dead water level operating zones, and shows no trend towards returning to the normal operating zone, the water level anomaly model is automatically paired up, and inter-station load is redistributed so that the abnormal water level can return to the normal operating zone as soon as possible, smoothing out the time accumulation effect caused by the calculation error. Specifically: For run-of-river hydropower stations, a water level control range Z_{down} to Z_{up} is fixed between the dead water level Z_s and the normal water level Z_x. When the real-time reservoir level Z_t satisfies $Z_{up} < Z_t \leq Z_x$ or $Z_s \leq Z_t < Z_{down}$, it is considered to enter an abnormal operation zone of high or low water level. If $Z_{down} \leq Z_t \leq Z_{up}$, it is considered to be in the normal operating zone. The principle of water level zoned control is shown in Fig. 5.3.

If the watershed reservoir levels are all in the operable zone and one or more power stations have water spillage, then the distribution model for minimum water spillage is automatically paired up to reduce the amount of electricity lost from the power stations by spillage.

If there is no spillage at any watershed reservoir, either the large-load distribution strategy or the small-load distribution strategy is adopted according to the magnitude of change of the overall generation load command value and of the actual total output contribution. Under command mode, four types of economic dispatch model are available, namely maximum energy storage, stable water level, less load adjustment, and load balancing. In the non-command mode, when conditions are met, automatic entry into either the water level anomaly model or the minimum water spillage distribution model is triggered, where the water level anomaly model has higher priority than the minimum water spillage distribution model.

In addition, under spot trading market mode, the cascade hydropower plants execute the spot trading results. The load distribution system obtains and executes the spot trading results, the system providing prompt warning when water level safety constraints might be violated.

5.2.2 Late Flood Season Staged Water Impoundment Technology for Regulating Reservoirs

The Dadu River Company has done much exploration in the scientific utilization of flood season flood water. Of particular reference value is the use of late flood season staged water impoundment for power generation dispatch for regulating reservoirs.

From analysis of the watershed's climate background for precipitation, atmospheric circulation conditions, water vapor sources, and flooding emergence timing and magnitude, the Dadu River watershed shows a clear seasonal change pattern, the

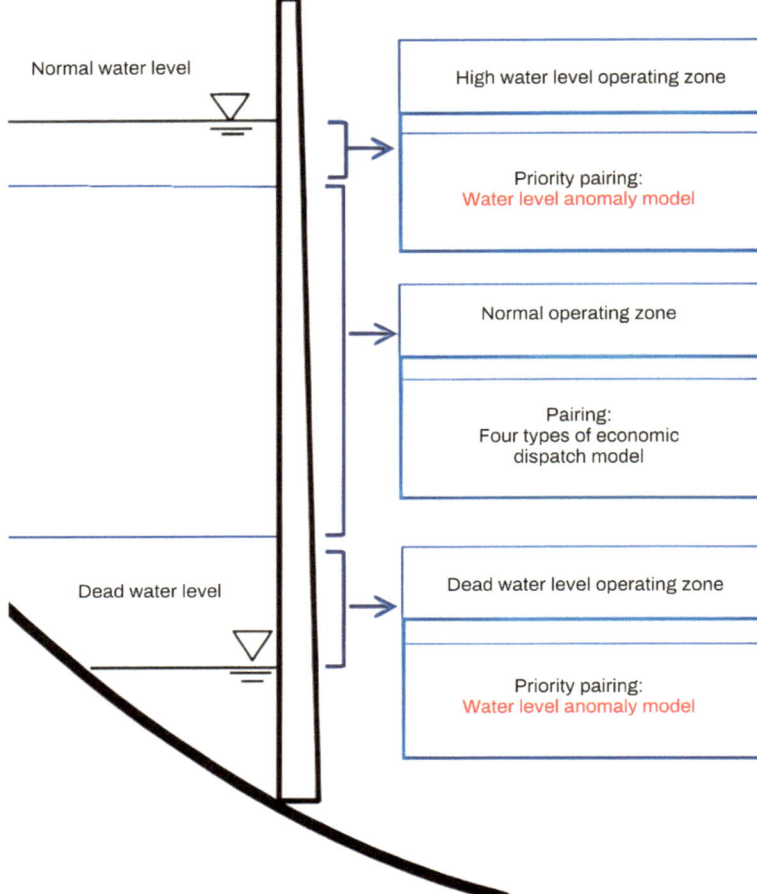

Fig. 5.3 Principle of water level zoned control

overall flooding process going from weak to strong, and then from strong to weak again. That is to say, the reservoirs go through a gradual process from non-flood period to flood period and then again from flood period to non-flood period. Based on mathematical statistics and system engineering, the Dadu River Company has continued exploration and has broken through the traditional medium- and long-term dispatch mode for flood-limited water level control, using methods such as flood period staged fuzzy analysis, variable point analysis, relative frequency, and circular distribution, combining staged reservoir water level control with medium and long-term marketing dynamics, and extending the cutting edge application domain of medium- and long-term optimal dispatch technology for cascade reservoirs.

While ensuring no increase of flood control risk, in order to improve the utilization efficiency of water resources, and to further put into play the storage and regulation role of Dadu River Pubugou reservoir, late flood season flood water entering the reservoir can be impounded at the appropriate time and to an appropriate degree. The key to the problem lies in two points: First is how to master the uncertainty of the reservoir water inflows, more accurately delineating the late flood season flood waters; Second is how to formulate a more scientific late flood season staged water impoundment solution to achieve a balance of risk and benefit.

Fuzzy hydrology sees the "flood period" as a fuzzy concept, the intermediate transition being the flood period's fuzziness, and the basis for fuzzy analysis. In the transitional stage, the stage the reservoir is in has mixed properties, to a certain extent of being both in a flood period and not in a flood period, and this is the scientific basis for fuzzy analysis of the flood period. The following is an example from the Pubugou power station reservoir, a key regulating reservoir for the middle section of the Dadu River watershed.

Based on actual hydrometeorological conditions in the watershed above the Pubugou power station, given the interval value of flood period physical cause indicators, when the entering-flood (or leaving-flood) indicator ($Q = 2500$ m^3/s) is less than the lower limit of the interval, the flood degree of membership is 0. When the indicator value is greater than the upper limit of the interval, the flood degree of membership is 1. According to the number of times from June to September that a day falls within the sampling interval for the flood year, we can get membership degree of time t belonging to the flood period, and the degree of membership of each day belonging to flood period shows as a single peak. If we take the membership degree of 0.80 or more as the cutoff to determine the main flood period, we can classify the main flood period as June 29 to August 12. However, in consideration of the random fluctuation of flood stage, the main flood period can be designated as late June to mid-August. The Pubugou reservoir flood period membership function curve is shown in Fig. 5.4.

From the scatter diagram for Pubugou reservoir cross-sectional peak flood flow (shown in Fig. 5.5), it can be seen that the August 10 to August 20 period is an obvious weak gap, during which annual maximum flood peak flows appeared relatively infrequently, with peaks and volumes increasing significantly afterwards. Looking at the ranking of maximum flood water annually at Pubugou reservoir, flood water flow volume after August 20 is greater than 5000 m^3/s magnitude for a total of only 3 times. For other floods occurring after August 20, the magnitudes are all less than 5000 m^3/s. Based on this, combined with the conclusion of the fuzzy analysis method above, the flood waters after August 20 can be considered as belonging to the flood transition phase from the main flood period to the dry period, with August 20 being the appropriate cut-off point between the main flood period and the transition period.

Based on the flood stage delineation results, multiple storage solutions are formulated for benefit and risk comparison. Since Pubugou reservoir still needs to reserve some reservoir capacity for flood control in August for the middle and lower reaches of the Changjiang River, so flood regulation calculations are carried out combining in the various constraints for reservoir dispatch, and a solution formulated to start storing water from September.

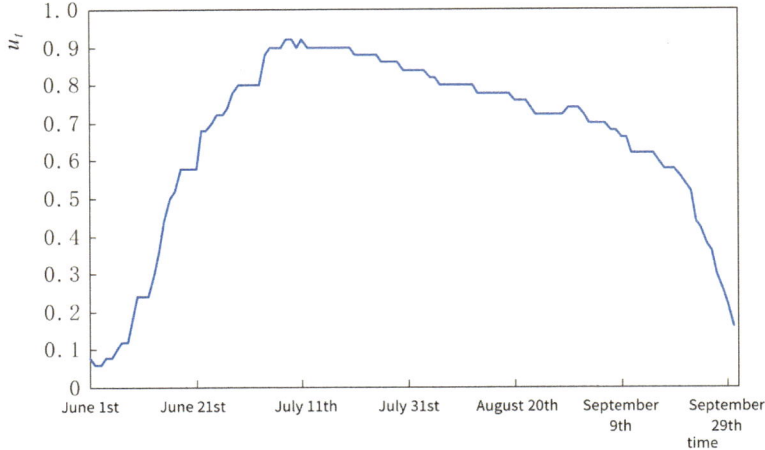

Fig. 5.4 Pubugou reservoir flood period fuzzy membership function curve

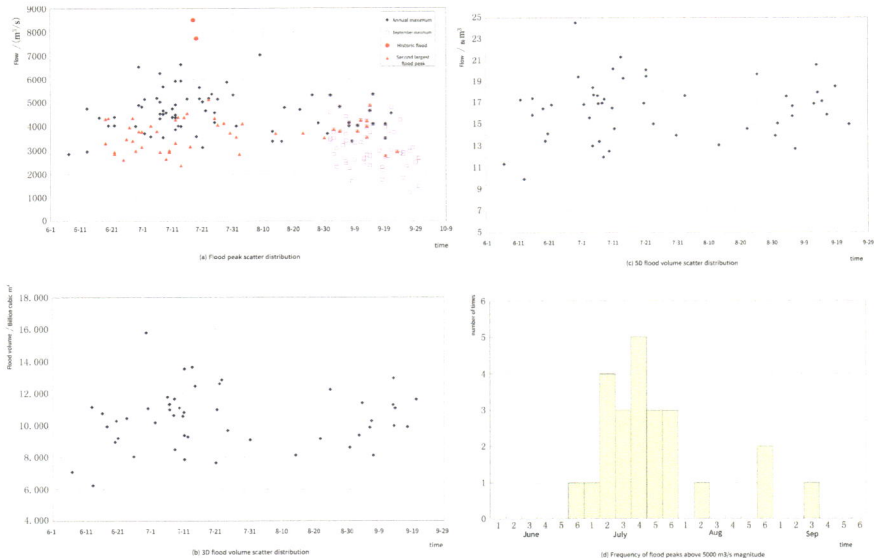

Fig. 5.5 Statistical analysis of Pubugou reservoir flood waters

In the original reservoir dispatch chart, the water level control plan for September to October was modified according to each storage solution. Benefit calculations for Pubugou with ten-day runoff series for a total of 70 years, from June 1937 to May 2007, found that starting storage on September 1 increased power generation the most, but reservoir water level after flood regulation was also greatest. If start of storage is only moved up to September 20, the additional power generation is too small, and this advance storage is not very meaningful. Generally, considering the difference in flood magnitude and flood causes in different periods of September, regardless of flood magnitude or occurrence frequency, there is clear distinction around September 15. September 15 is a more obvious inflection point, thus starting water storage on September 15 is a better solution. Additional power generated is 0.51 billion kWh, which is only 0.16 billion kWh less than starting water storage September 10, but 0.35 billion kWh more than starting September 20. However, the highest reservoir level for check flood water level after flood regulation is not much different from the solution with water storage starting on September 20, which basically achieves the balance between flood prevention risk and additional power generation benefit.

5.2.3 Spot Market Marginal Clearing Price Forecasting Technology

The term "marginal price" refers to: During spot power market power trade, power is traded in the order of lowest to highest price one by one, such that the last power supplier's price to meet the load demand is called the marginal price of the system. China's Sichuan power market employs a unified clearing price model, so the marginal power price is crucial for the spot market, not only determining the overall market transaction price, which is settled in terms of the marginal price, but also affecting the power quantities traded by each market actor. However, the Sichuan power market is characterized by a high proportion of installed hydropower, complex power dispatch relationships, and large disparities in generation costs. Based on the analysis of the factors influencing the marginal clearing price, the Dadu River Company built a support vector machine (SVM) to build a forecasting model, using an improved evolutionary algorithm based on population-based incremental learning—the DPBIL algorithm carries out optimization of SVM's penalty parameter C and kernel function parameter g, to improve the extendibility and generalizability of SVM, forming the DPBIL-SVM hybrid forecasting model. The structure of the DPBIL-SVM hybrid algorithm is shown in Fig. 5.6.

Based on availability of actual measurement information, training sequences and test sequences for short-term electricity price forecasting are selected from factors such as the results of electricity supply and demand analysis, weather, and daily classification as model inputs, and forecast results as model outputs. The model learns the training sequences, determines the forecast model parameters, adopts decimal coding to address the issue of code redundancy, uses the principle of equal probability of alleles to determine the initial probability, and uses the cumulative probability roulette wheel selection method to generate the initial population. For the probability conflict problem, improvement to the evolutionary approach is by adding a correction

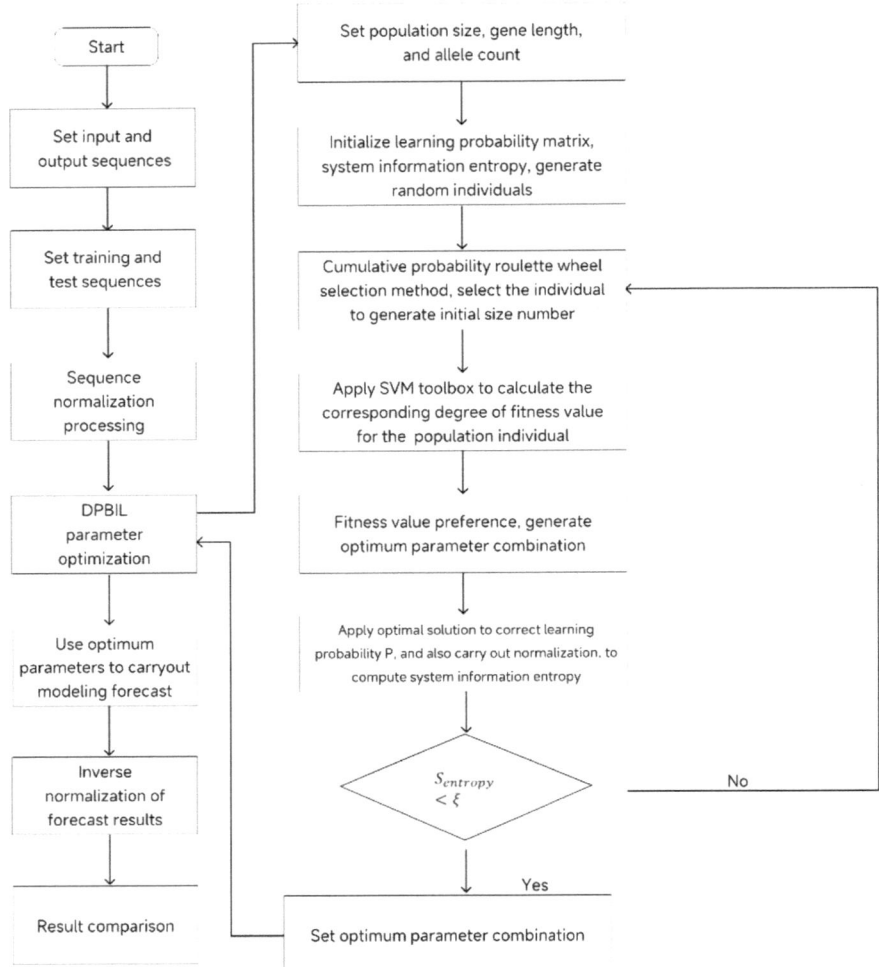

Fig. 5.6 DPBIL-SVM clearing price forecast model structure

factor x to the allelic probability corresponding to the optimal solution, and carrying out normalization of all probabilities.

The DPBIL-SVM hybrid algorithm is characterized by both overall evolution and global optimization. In one evolution process, sixty parameter pairs are selected, and each parameter pair is tested 432 times. The domain of convergence of system information entropy is reached after about 1000 evolutions. Data processing efficiency is more efficient than single SVM exhaustive enumeration, and sensitive information is well captured. Comparisons of training and test sets are shown in Fig. 5.7.

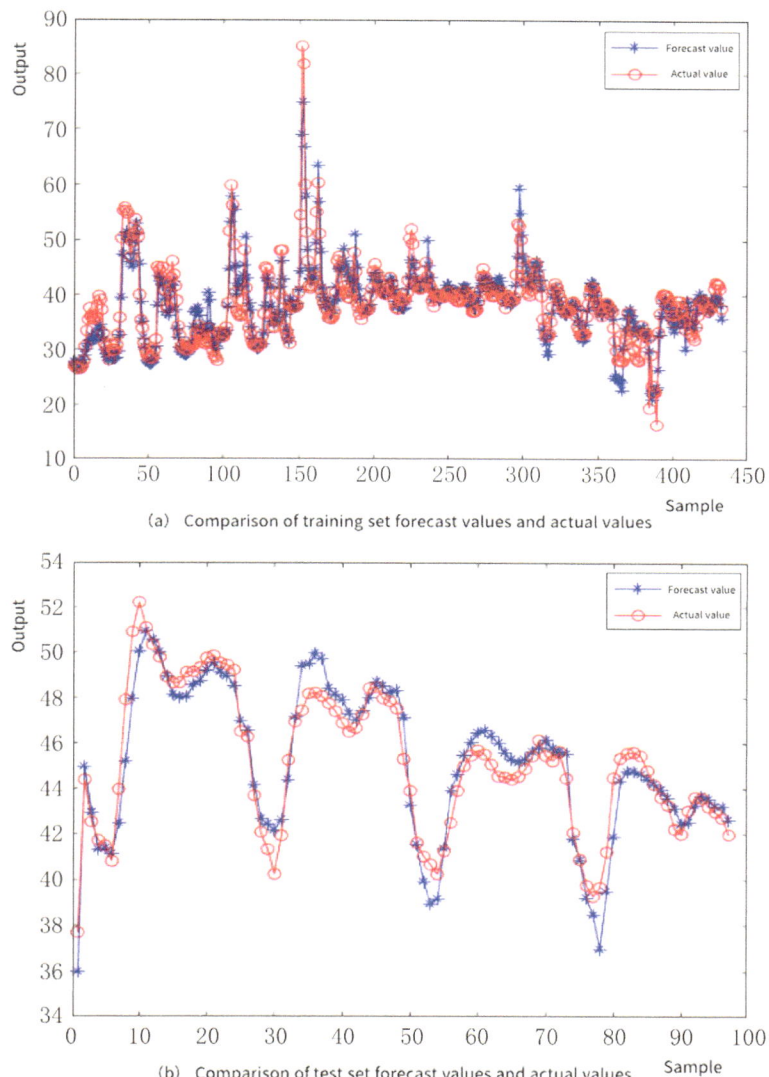

(a) Comparison of training set forecast values and actual values

(b) Comparison of test set forecast values and actual values

Fig. 5.7 DPBIL-SVM training and test set results

5.2.4 Spot Trade Decision-Making Support Technology

Trading decision-making for the Dadu River cascade hydropower power market is gradually launched around different time series. On the basis of situational analysis of medium and long-term supply and demand, and forecasting of equipment health and water inflow, medium- and long-term and spot trading are carried out on a rolling basis to rationally control the water level of regulating reservoirs, ensure contract execution, and strive to maximize medium- and long-term power generation benefits. At the same time, through scientific formulation of offering strategies in the spot market, the execution of medium- and long-term contracts is ensured, along with safe and economic production operation as well. The organic interconnecting and rolling optimization of long-, medium-, and short-term and spot trading strategies, lead to the overall optimization of hydropower energy resources over a relatively long time cycle, achieving the safe and reliable operation of cascade hydropower plants, and overall optimal benefit.

(1) Approach to Regional Spot Trade Decision-Making Support

Spot trading in China's Sichuan power grid, where the Dadu River watershed is located, includes both day-ahead and real-time spot trading. In day-ahead trading, hydropower companies are required to declare their 96-point, ten-segment, power output and electricity pricing for the next day. In real-time trading, companies are required to declare their power generation capacity for the next few hours one hour in advance. In combination with Sichuan spot trading rules, the Dadu River Company declares generation output along the following lines:

(2) Approach to Day-ahead Trading Output Declaration

First, according to medium- and long-term water inflow forecast and price trends, medium- and long-term optimal dispatch for the cascade is developed, with medium- and long-term optimal dispatch water level control strategy as the boundary condition for measurement and calculation of the spot trading output process. Then, next day water inflow conditions and marginal clearing price are forecast and treated as input conditions of the spot trading output process measurement and calculation.

Secondly, under the given boundary conditions and input conditions, short-term optimal dispatch for the cascade is developed to obtain the future 96-point cascade generation output process. Finally, according to the trading rules, day-ahead 96-point, ten-segment, output and pricing strategy is obtained via adjustments according to various principles.

(3) Approach to Real-time Trading Output Declaration

As there exist many factors that change during the actual intraday production process, such as changes in water inflow from upstream or between reservoirs, changes in the power output processes of upstream power stations, changes in equipment, and changes in market environment, etc., the impact of changes can be corrected through real-time trading. Based on the latest water conditions and market information, forecasts of water inflow and marginal outgoing price are carried out, boundary conditions are modified, rolling development of cascade short-term optimized dispatch is done, and the output declaration solution is optimally adjusted.

(4) Spot Trade Output Declaration Model

In day-ahead and real-time spot trading, each hydropower station on the Dadu River needs to declare 96 points of output for the next day or for several hours in the future.

This requires scientific preparation of power generation plan through short-term optimal cascade dispatch.

The optimized dispatch model considers constraints such as the power grid, cascade water level limits, water volume linkages, rough zones, plans for servicing, etc., and the power generation output of the cascade hydropower plants is fixed with the goal of maximizing power generation revenue according to determinations of water level control strategy, forecast runoff, and marginal clearing price for medium- and long-term optimal dispatch of the regulating reservoir. Day-ahead spot trading is calculated for day periods, to obtain a 96-point power generation output process. Real-time spot trading is calculated on a rolling basis for periods covering the next few hours. The output declaration model is:

$$
or \quad \left.\begin{array}{c} I = \text{Max} \sum\limits_{i=1}^{N} \sum\limits_{t=1}^{T} \left(k_i \cdot Q_{i,t} \cdot H_{i,t} \cdot M_t \cdot p_t \right) \\[3mm] I = \text{Max} \sum\limits_{i=1}^{N} \sum\limits_{t=1}^{T} \frac{Q_{i,t} \cdot M_t \cdot p_t}{\delta_{(i,t)}} \end{array}\right\} \tag{5.1}
$$

The model also must take into account water volume balance, hydraulic connections between cascade hydropower stations, power output capacity, the water passing capacity of each hydropower plant generator set, and other constraints (the specific formula will not be elaborated).

In the formula above:

k_i is the output coefficient of power plant No. i;

$Q_{i,t}$ is the power generating volume (m^3/s) of power plant No. i during time segment t;

$H_{i,t}$ is average net water head (m) for power generation of power plant No. i during time segment t;

M_t is number of hours in time segment t;

$\delta_{(i,t)}$ is water consumption rate of power plant No. i during time segment t;

p_t is the electricity price factor for time segment t;

T is the total number of calculated time segments in the dispatch period;

N is the total number of cascade power stations.

- Spot Trade Optimization Algorithm

Cascade hydropower short-term optimal dispatch and optimal load distribution is a multi-state programming problem, under a variety of complex nonlinear constraints. Increase in the number of power plants and the number of computational time segments, along with increases in accuracy, will cause significant increases in the computational dimensionality, which can easily lead to the "curse of dimensionality" problem. In order to seek a balance between solution efficiency and quality, the Dadu River watershed's optimization calculation employs the progressive optimality algorithm (POA).

The progressive optimality algorithm was proposed in 1975 by the Canadian scholars H. R. Howson and N. G. F. Sancho, for use in seeking out solutions to multi-state dynamic programming problems. The algorithm proposes the principle of progressive optimality based on the idea of Bellman optimization, i.e., the optimal route has the property that each pair of decision sets is optimal with respect to its initial trajectory value and termination value. The algorithm decomposes the multi-stage problem into multiple two-stage problems, and solving the two-stage problem is just a search for optimality for the decision variables of the selected two stages, while fixing the variables of the other stages. After resolving the problem for the current stages, the next two stages are then considered, using the last result as the initial condition for the next search for optimality, and repeating this cycle until convergence.

During the short-term power generation capacity optimization calculation, the dispatch period of the cascade power station is one day, and the dispatch time interval is 15 min, so the whole dispatch period is broken into 96 time periods, the number of cascade power stations is N, and the power station No. is i ($0 \leq i \leq N$-1). The algorithm flow is shown in Fig. 5.8.

5.2.5 Technology for Checking the Spot Market Clearing Result

The complex hydraulic and electric power interconnections between cascade power stations, uncertainty of water inflow, and fluctuation of electricity prices have, to a certain extent, increased the complexities for participation in the power market by hydropower. Additionally, the power output of cascade power stations cleared by the market can easily be mismatched to the generator sets and fall in their rough zones, so the clearing results may lead the enterprise to face potential safety or economic risks.

The Dadu River Company uses the basic methods and theories of hydroelectric station water power calculation, and considers the multidimensional and nonlinear constraints in the process of operating power plants, to simulate to the maximum extent possible the processes of reservoir water level operation and power generation output by generator sets, and to discover in advance the risks that the clearing results may create in regard to the safe production and economic operation of the hydropower plant.

Checking technology for spot market clearing results starts predominantly from the power station itself. Simulated projection of the load process following market clearing is carried out, considering the multi-dimensional, non-linear constraints in the hydropower operation process, and reconstructing the hydropower station water level and generator set output process for power production based on the load process. Thorough investigation is carried out as to possible risks of causing reservoir water level anomalies, long-term rough zone operation of generator sets, or violation of overall principles for utilization of water resources. Estimates of power loss generated as a consequence of water spillage are made, along with assessments of the benefits

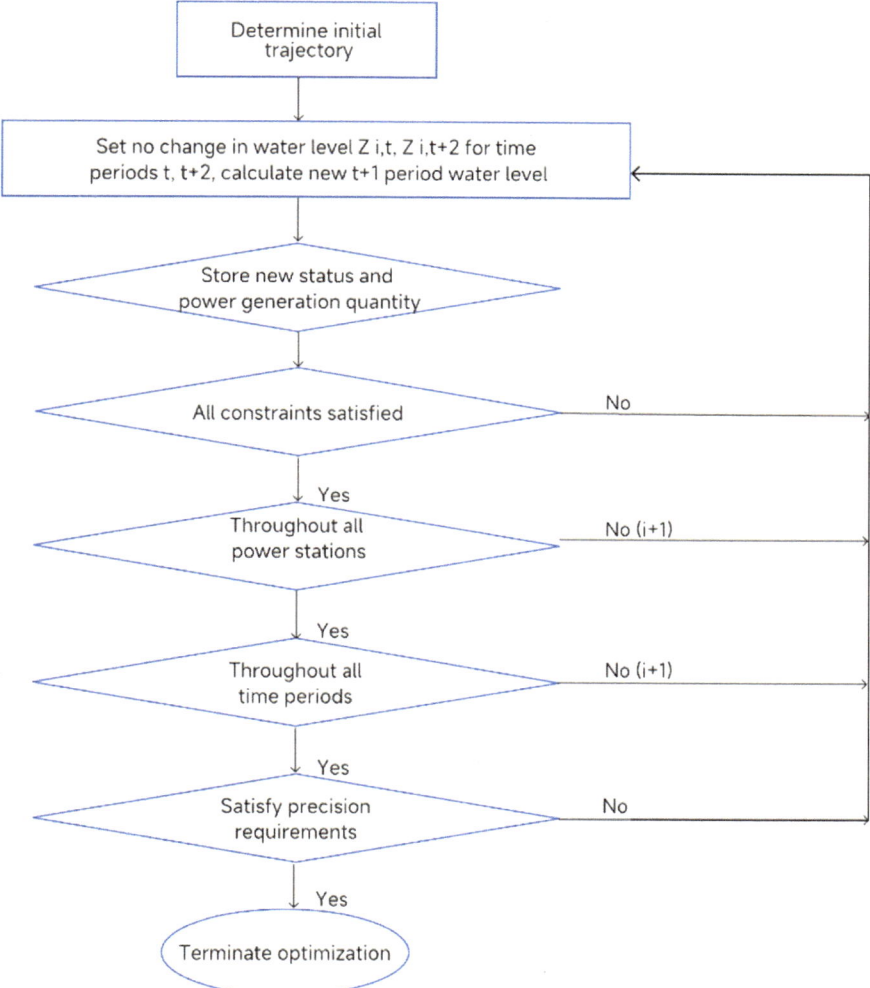

Fig. 5.8 Progressive optimality algorithm flow

generated by this bid offering. Successful or unsuccessful bids are analyzed and summarized as model bidding cases. When fairly large risks exist, corrections can be made via the real-time market in order to ensure maximization of safe production and economic benefits.

Power market clearing result checking technology takes into consideration a wide range of complex nonlinear constraints related to hydropower station operation process, such of water volume balance constraints, hydraulic connections between cascade power stations, reservoir water level and discharge flow constraints, power station generator set water flow volume and rough zone constraints, and power station output ramp limit constraints, along with the influence of flow propagation time between the cascade power stations on the optimal operation of power stations. Then, on the foundation of original hydroelectric plant water power calculation theories and methods, a chunked multiple time sequence constant output algorithm is designed to reconstruct the reservoir power plant operation output water level process. Based on this, evaluation indicators are selected, and analysis is carried out on the economics of the offering, predicting water spillage in the process of power plant production, and measuring and calculating the benefits of a single complete offering instance in terms of "contract for difference" (CFD), guiding the power plant, from multiple dimensions, to make load adjustments or real-time power volume declarations.

Combining in any generator set servicing and adjustment shutdown plans, etc., for that day, the number of usable generator sets in the power station is determined, and corresponding power station rough zone range checks for the corresponding power stations are calculated. After determining the clearing result and initial water level for the power station, the system carries out the water level process check calculations, combining in water volume balance and the influence of flow propagation time between the cascade power stations, and considering the numerous power station constraint conditions, obtaining a 96-point reservoir water level process for the check day, and carrying out zoning for reservoir water level ranges. When the clearing result puts the reservoir into low water level operation, a timely warning is made. According to power station water level and flow requirements for flood control, water supply, shipping, and maintenance of ecological stability, overall usage checks are performed, and early warning is made when the flow or water level calculation results do not satisfy requirements. For power stations where water spillage occurs, time periods when spillage occurs are selected out and used to statistically calculate the volume of generated spillage. The results of medium- and long-term contract energy decomposition, and the completion status for medium- and long-term contract energy, are checked, and the benefits of spot trading are calculated based on the contract for difference method: Using the contract for difference formula, clearing result benefits are calculated for the results of a single trade instance.

In the environment of the spot market, power production is guided by power marketing results, i.e., the actual power generation process for power stations should be carried out in accordance with the post-trade generation load curve. However, during the actual power generation process, due to the impacts of a series of factors such as the actual working conditions, generator set operating circumstances, AGC load adjustment, etc., the actual power generation process will see certain divergences between the declaration curve and the trade curve, destroying the hydraulic and power matching relationships of the original declaration and trade results between upstream and downstream cascade power stations. Therefore, the actual power generation process is tracked in real-time, and compared with the declaration and trade curves, then the power generation process is adjusted to fit the trade curve process as much as possible. The reasons for the disparities that emerge during actual power generation process should be promptly analyzed and corrective measures taken.

5.3 Case Application

5.3.1 The First Time to Achieve Power Generation Load Smart Cooperative Dispatch for a Large-Scale Watershed Cascade

5.3.1.1 Application Background

By the end of 2018, hydropower, solar, wind, and other renewable energy in Sichuan had a full installed capacity of 98.327 million kW, representing 87.4% of the province's total installed capacity. Affected by the uncertainty of water inflows and wind movement, and the poor water regulation capacity of the vast majority of hydropower in Sichuan, the load on Sichuan's peak and frequency regulating power stations varies greatly and adjustments are frequent, with great randomness and uncertainty. In June 2019, the southwest China power grid was asynchronously interconnected with an external grid outside the region, the external grid having a synchronous grid size of only about 1/6 of the original southwest-central-north China grid, and this in particular further highlighted the structural contradictions of the power supply of the southwest grid, and created even greater response speed requirements for load adjustment by generator sets regulating peak and frequency. The three hydropower stations in the middle and lower reaches of the Dadu River, namely Pubugou, Shenxigou, and Zhentouba, with a total installed capacity of 4.98 million kW, form a large-scale hydropower station group that is the main force for peak and frequency regulation for the Sichuan power grid. In actual operation, the reservoir storage volumes of Shenxigou and Zhentouba are small, water level is impacted by load upstream at Pubugou, and AGC at the conventional power station level has difficulty controlling water level in a coordinated and economic manner. It often has to be deactivated and can only passively accept power grid dispatch "fixed load" commands. This leads to the two reservoirs being highly prone to large fluctuations in water level, water spillage, and reservoir emptying, seriously affecting the safe and economic operation of the cascade power stations.

In April 2017, the cascade power station real-time smart one-button load dispatch system (one-button load dispatch) developed by Dadu River Company was successfully put into use, realizing real-time and smart load distribution between cascade power stations. After being put into use, work efficiency, speed and accuracy of load regulation, power station operation efficiency, and utilization of water potential, all saw significant improvements.

5.3.1.2 Load Adjustment Scenario

The Pubugou power station has a regulating reservoir capacity of 3.826 billion cubic meters, with incomplete annual regulating capacity. Downstream, Shenxigou and

Zhentouba (Level I) are the re-regulating power stations for Pubugou. Shenxigou power station has a regulating reservoir capacity of only 8.48 million cubic meters, and Zhentouba (Level I) hydropower station has a regulating reservoir capacity of only 12.3 million cubic meters, so reservoir water levels are considerably impacted by Pubugou load changes. In particular, the Pubugou power station has been directly receiving active commands from the grid AGC for a long period, and load fluctuates frequently, so operation duty personnel need to keep a close eye on dozens of parameters, including Pubugou load changes and resultant changes in reservoir levels at the Shenxigou and Zhentouba (Level I) power stations, rainfall status information for the watershed, and fluctuation of transmission cut-planes. Personnel then must manually calculate the water volumes and load matching results for the three stations in real-time. In cases of poor load matching, higher-level dispatch personnel must be contacted in time to apply for corresponding load adjustments so that the load of the two downstream stations can be matched in real-time with Pubugou as much as possible, avoiding large up and down fluctuations in reservoir water levels, and reducing occurrences of water spillage or reservoir emptying.

Conventional manual load adjustment scenario: On a certain day, the water level of Shenxigou power station is 658.5 m (normal water level 660 m), with 1800 m^3/s inflow, 1500 m^3/s outflow, with a flow difference of 300 m^3/s, and 45 cm/h water level fluctuation. Pubugou power station, as the primary peak and frequency regulating power station for the Sichuan power grid, accepts automatically issued load commands from the grid AGC, and Shenxigou power station accepts commands from grid duty personnel, as arranged by the grid. The water level of Zhentouba (Level I) hydropower station is 621.5 m (normal water level of 623 m). Its reservoir inflows and outflows are in balance, and thus, the water level is stable. However, Shenxigou, the upstream power station, is bound to disrupt this balance once its load increases.

Affected by the continuous increase of load during power grid peak hours, the load for Pubugou power station continues to increase, but there is no attendant command for the Shenxigou power station. The operation duty personnel contact the grid several times wanting to increase the load at Shenxigou by 200 MW, so as to control the swelling water levels. But power grid dispatch does not consent. The water level at Shenxigou quickly rises to 659.1 m, close to the water level upper limit of 660 m. The water level continues to rise rapidly, seriously endangering the safety of the reservoir. At this time, power grid dispatch is again contacted to apply for an increase in load of 300 MW for Shenxigou power station and 100 MW for Zhentouba (Level I) power station. However, the power grid only agrees to add 200 MW to the Shenxigou power station for the time being.

After receiving the load adjustment command, operation duty personnel quickly manually adjust the load of each operating generator set in turn, according to the required ramp rate of 60 MW per minute, and adjust the on-grid output of the power station to the target value. At the same time, personnel strictly monitor the rising water level, continue to appeal to provincial dispatch to increase the load, and make preparations to activate flood discharge facilities in order to prevent major safety accidents such as water inundating the dam or flooding plant buildings.

According to statistics, under the conventional manual load adjustment mode, the three stations of Pubugou, Shenxigou, and Zhentouba (Level I) applied to grid dispatch for load adjustment on average more than 30 times a day, generator sets started and stopped more than 10 times, and load was manually adjusted more than 150 times.

In order to avoid missed or late adjustment giving rise to energy deviation penalties, the operation duty officer needs to set up regular timed reminders. In addition, operation duty personnel must simultaneously consider generator set operation rough zones, load adjustment ramp rate, cascade spillage, and economic operation of the plant, with high work intensity and elevated pressure while on duty.

Conventional one-button load dispatch: With implementation of real-time smart one-button load dispatch system for cascade hydropower stations, the grid AGC issues the total load command for Pubugou, Shenxigou, and Zhentouba to the cascade real-time smart one-button load dispatch system according to system load demands or frequency change circumstances. The cascade real-time smart one-button load dispatch system then smartly distributes the total load according to the control strategy, and sends it to the three power station AGCs, which individually execute the distributions and return feedback on execution results.

A typical adjustment scenario is as follows: The cascade real-time smart one-button load dispatch system is put into "provincial dispatch" operation, and dispatch issues a given value of total active power for Pubugou, Shenxigou, and Zhentouba of 3000 MW, up from 2800 MW. After receiving the grid command, the one-button load dispatch system automatically selects up a load distribution strategy for stable water level, compares that against various safety constraints, and speedily gives distribution results as follows: Pubugou load is set from 1980 to 2130 MW, Shenxigou load is set from 420 to 450 MW, and Zhentouba load is set from 400 to 420 MW, and this configuration is automatically immediately issued to the power station AGCs to execute the load adjustments. During operation, if there is a possibility of water level at Shenxigou or Zhentouba exceeding the normal operating range, without showing a trend toward return, then the one-button load dispatch system will automatically detect this and make a determination to redistribute the inter-plant load, adjusting the load so that, as far as possible, reservoir levels will operate within a reasonable range.

With implementation of the one-button load system, power station load is smartly adjusted, reservoir water level is automatically optimally controlled, and the whole adjustment process is completely free from human intervention. Duty personnel no longer keep their eyes ceaselessly fixed on data such as reservoir water levels, reservoir inflows and outflows, power system flows, etc. Annually, this has reduced the number of manual load adjustments at Shenxigou and Zhentouba (Level I) power stations by more than 30,000 times, and has effectively decreased work intensity for dispatch personnel. At the same time, through the logic of smart strategy selection based on water level operating zones, safety is improved for fields such as hydraulic engineering and flood prevention. Through dispatch strategies such as less load adjustment and stable water level, and through the dispatching strategy of maximum energy storage and minimum water spillage volume, operational economy for cascade hydropower plants is effectually improved.

For the grid, implementation of cascade real-time smart regulation technology for load has altered the real-time regulation model of the Sichuan power system, transforming from a two-tier regulation model of "grid to power station," to a three-tier regulation model of "grid to cascade regulation to power station." Additionally, multiple stations are bound together to participate jointly in grid peak regulation, which increases the grid peak regulating capacity and achieves a win–win situation for both the power stations and the grid.

5.3.2 The Clear Economic and Social Benefits of the Optimized Late Flood Season Water Impoundment Process

The value of reservoir regulating capacity is that it can effectively store and regulate water inflows, and improve the utilization rate of water potential and the quality of the electricity supply. The main regulating reservoirs on the Dadu River bear responsibility for downstream flood control, and these dams need to set aside a certain amount of reservoir capacity for flood control, affecting exercise of the storage and regulation functions of the reservoirs. For this reason, it is necessary to achieve refined dispatch of the cascade through technologies such as late flood season staged water impoundment. Both the *Changjiang River Basin Flood Control Planning Report* and the *Changjiang River Basin Comprehensive Utilization Planning Report* require Pubugou to reserve reservoir capacity in August for downstream flood control and also in August to undertake the task of downstream flood control. Therefore, for Pubugou reservoir staged water impoundment, advance water impoundment in September is fitting.

In July 2016, high-intensity sustained rainfall failed to form on the upper reaches of the Dadu River watershed, and average reservoir inflow at Pubugou, the key regulating reservoir for the middle reaches, was 2270 m³/s, down 12.25% from the previous year, and 10% drier than in many years. In August, the Sichuan Basin continued to be affected by high temperatures and summer drought conditions. Average reservoir inflow at Pubugou was 1650 m³/s, about equal to the same period the previous year, and 20% drier than in many years. Based on the inflow situation for the primary flood season of July to August, and employing regulating reservoir late flood season staged water impoundment technology, the Dadu River Company carried out late flood season water impoundment analysis, and set September 15 as the start for Pubugou reservoir late flood season water impoundment.

In early September 2016, the Pubugou reservoir started gradual impoundment towards the flood-limited water level. After September 15, water impoundment efforts were increased. On September 30, Pubugou reservoir reached a final impoundment level of 847.07 m. In September, Pubugou reservoir water resource utilization rate reached 100%.

In 2016, the Dadu River Company used late flood season staged water impoundment technology to detain an extra 481 million cubic meters of water, which translates into 336 million kWh of electricity overall at Pubugou plus the lower three cascades stations of Shenxigou, Gongzui, and Tongjiezi, saving 115,900 tons of coal consumption for power generation, reducing soot emissions by 78,800 tons, and reducing greenhouse gas emissions by 301,900 tons.

Since 2013, through application of late flood season staged water impoundment technology, by the end of September each year Pubugou reservoir water storage reaches about 846 m (flood-limited water level is 841 m). As of the end of 2020, extra dry-period power generation at Pubugou and the downstream hydropower stations belonging to the Dadu River Company was cumulatively about 2.022 billion kWh.

The implementation of late flood season staged water impoundment can significantly reduce the impact of reservoir water impoundment during normal water level periods on downstream cascade power generation output, and improve the electricity supply capacity of the Sichuan power grid during normal and dry low water level periods. The extra hydropower generation will reduce carbon emissions pressures and make positive contributions to the construction of a resource saving and environmentally friendly society in Sichuan Province.

5.3.3 Successful Support of Cascade Power Station Spot Trade During Trial of Spot Trading in Sichuan

Sichuan Province is one of eight initial pilot spot market areas confirmed in China. Sichuan Province launched the construction of the power spot market in 2017, in order to implement the arrangements of the National Development and Reform Commission, the National Energy Administration, and the Sichuan Provincial Government regarding the work of construction of the pilot spot market for power. As the construction progressed, Sichuan Province organized simulated trading, power generation dispatch trial runs, and settlement trial runs, respectively. Simulated trading means that market actors conduct spot trading using the trading platform, but do not actually dispatch power generation or settle up based on the clearing results. Power dispatch trial run means that the power grid dispatches power generation to power generating enterprises based on the clearing results, but does not settle on this basis. Settlement trial run means that the power grid dispatches power based on the clearing results, and uses the clearing results and power generation situation as basis for settlement.

From September 26, 2020, to October 25, 2020 (i.e. October "Power Market Month"), Sichuan Province conducted a one-month trial run of spot trading settlement. This period coincided with the transition from a period of abundant water, to a period of normal and dry low water level in Sichuan, with water inflows from the watershed falling rapidly. The regulating reservoirs urgently needed to be filled up, and the effective power generation capacity of each power station was falling rapidly. The market supply and demand situation rapidly evolved successively from supply exceeding demand, then to basically balanced, and finally to demand exceeding supply. This created the following requirements for the marketing team:

(1) The water inflow process must be accurately forecast, and precise predictions made of the daily power generation capacity of each power station belonging to the company.
(2) Changes in the province-wide supply and demand relationship must be accurately researched and judged.
(3) The patterns of price change for each time interval each day must be fully understood.
(4) An offering strategy for each power station belonging to the company must be scientifically formulated.

The Sichuan spot market employs a centralized market model, and the system marginal price is used for price clearing. Priority power generation plans and outgoing power volume serve as the boundary of the spot market, and are physically executed. Medium- and long-term power contracts within the province are based on contract for difference. Wholesale market users and power sales companies quote volume but do not quote price in the day-ahead market, and use contract for difference to settle according to the actual power consumption curve, the day-ahead demand curve, and the medium- and long-term settlement curves.

In October 2020, the Dadu River Company used spot trading decision-making support technology in areas such as medium- and long-term contract power decomposition, supply and demand research, marginal clearing price prediction, trading decision-making support, and clearing result checking, to achieve efficient, scientific, and optimized spot trading, which effectually enhanced the market competitiveness of the enterprise.

(1) Energy Decomposition in Medium- and Long-term Cascade Contracts

At present, there are numerous varieties of medium- and long-term contracted power in Sichuan Province, including primarily: inter-province and inter-region power, priority power, annual and monthly (weekly) bilateral negotiated trade, annual and monthly (weekly) direct trade, annual and monthly (weekly) listed trade, surplus power, spillage period trough time power, targeted support power, generation rights, and contractual trade, etc. By means of the spot trading decision-making support system, decomposition is carried out for the medium- and long-term contracted power for the Dadu River Company's nine cascade power stations. Contracted power is decomposed into a daily 96-point output process according to market rules, and as contracted. This allows trading personnel at all times to clearly and conveniently grasp the status of the indicators for medium- and long-term contracted power of each power station, and as a priority confirmation task of the power declaration strategy, it avoids the loss of contracted power due to errors in trading strategy, and provides clear improvements to efficiency in making a declaration.

(2) Marginal Clearing Price Analysis and Forecast

Marginal clearing price analysis and forecast is the key in spot trading. In October 2020, with the launch of the power market in Sichuan Province, the Dadu River Company collected daily network-wide data on power generation, consumption, external purchasing, outgoing transmission, clearing price, and weather, and used them along with date type (whether it was a holiday or special event date) as impact factors, to build a marginal clearing price forecast model based on artificial intelligence algorithms such as support vector machines. However, because this was just during the early period of the spot trading market, the volume of accumulated historical data was fairly limited and temporarily unable to meet the needs for model training and electricity price forecast. In order to test the validity of the model, the Dadu River Company conducted a test of the model using historical data from the European spot trading market, with good results. At the same time, European market historical data was used to analyze the main factors impacting electricity price, and correlation analysis was carried out for Sichuan power market supply and demand conditions and electricity price, to guide spot trading declaration strategy. As historical data continue to increase, subsequent correlation analysis can be conducted on more impact factors to gain insight into some of the marginal clearing price patterns, continuously improve the model, and model training can be conducted to gradually improve the accuracy of marginal clearing price forecasting.

(2) Formulating Spot Trade Declaration Strategy

Following Sichuan spot market trading rules, a cascade volume-price declaration strategy optimization model is established, with the objective of maximizing overall effectiveness, according to boundary conditions such as watershed water conditions, available equipment capacity, marginal clearing price, reservoir target water level, and while fully considering the reservoir operation, equipment safety, reservoir shoreline stability, passageway congestion, and other constraints. According to the market supply and demand situation, a spot trading strategy is determined.

The Dadu River Company uses the spot trading decision-making support system to automatically generate a 96-point, ten-segment, volume-price declaration solution for each power station. In the Dadu River watershed, the month of October is generally a period of transition from flood to dry season. The first half of this period may be a period of spillage, and the second half may be a period of non-spillage. Different spot trading declaration strategies will be adopted for spillage and non-spillage periods. In 2020, September 26 to October 15 was a spillage period, when supply exceeded demand. The general spot declaration strategy revolved around greater power generation, less water spillage, and actively endeavoring for power generation targets. The days from October 16 to 25 was a non-spillage period, when demand exceeded supply. The general spot declaration strategy was to comprehensively consider water inflows and reservoir storage targets, and scientifically formulate a volume-price declaration strategy suited to upstream and downstream cascade power stations. Finally, the optimized offering solution was formed according to the offering strategy, guiding market trade. Going through the spot trading decision-making support system significantly improved the efficiency of spot trading, optimized the matching of power generation targets for cascade power stations, effectually reduced power lost through cascade water spillage, gave priority assurance of execution of medium- and long-term contracts, and tangibly improved the power generation effectiveness of the cascade hydropower stations.

- Cascade Power Station Trade Checking and Timely Correction

After market clearing, the Dadu River Company carries out simulated projection of the load process based on the market clearing results, considering the multi-dimensional and non-linear constraints in the process of hydropower operation, reconstructing the water level process when the reservoir is producing electricity according to the load process, isolating the risks that may cause reservoir water level anomalies, affect overall utilization requirements, or cause rough zone operation of generator sets, carrying out prediction of the amount of spillage power to be created, and examining of the effectiveness of the offering for the day. For clearing results with greater safety risks, timely appeal is made to the trade and dispatch centers. For mismatch between cascade tiers, power declaration is made again during real-time trade, and rolling matches made for cascade power generation load, to ensure the safety and effectiveness of the hydropower plants to the greatest extent.

During the spot trading process in October, 2020, in several instances, the system identified circumstances of load mismatch between the cascade power stations, and gave recommendations for cascade load adjustment. In particular, at three different times during the non-spillage period, warnings occurred for cascade load mismatch, and power adjustments were made during real-time trade, avoiding water resource wasting due to load mismatch.

The Dadu River Company spot trading decision-making support related technologies have better resolved the problems of market information perception, marginal clearing price prediction, optimization of cascade bid offering solutions, and bid clearing result checking, significantly improving the competitiveness and work efficiency of power generation enterprises participating in the spot market.

Chapter 6
Intelligent Equipment Operation and Inspection

6.1 System Architecture and Build Objectives

6.1.1 System Architecture

The Dadu River Company widely applies advanced technologies such as cloud computing, big-data, IoT, mobile internet, and artificial intelligence, integrates specialized technologies such as smart sensing and execution, smart control, and management decision-making, and incorporates advanced management concepts, to realize automation of information retrieval, networking of data transmission, smart analysis of data, and scientification of decision-making support.

Intelligent equipment operation and inspection in the Dadu River watershed is primarily accomplished through the building of platforms for information perception, operation control, data management, assessment and diagnosis, and decision-making command. The system architecture is shown in Fig. 6.1.

(1) The *information perception platform* carries out comprehensive perception of information related to power plant equipment, including generators, turbines, transformers, and auxiliary equipment, to dynamically comprehend equipment operation circumstances in real-time, carry out data governance according to data standards, and store data in the big-data platform.

(2) The *operation control platform* uses automation technology to increase the level of power station automation, reduce on-site operations by personnel, and gradually achieve destaffing on-site. For example, the platform accomplishes remote control of equipment through computer monitoring systems, and smart inspection of equipment via smart robots or smart inspection systems, and smart cooperation and linkage between monitoring, industrial television, and other business systems is achieved through establishment of a multi-system linkage model.

© The Author(s) 2025
Y. Tu, *Management of Hydropower Enterprises*, Water Resources Development and Management, https://doi.org/10.1007/978-981-97-5584-4_6

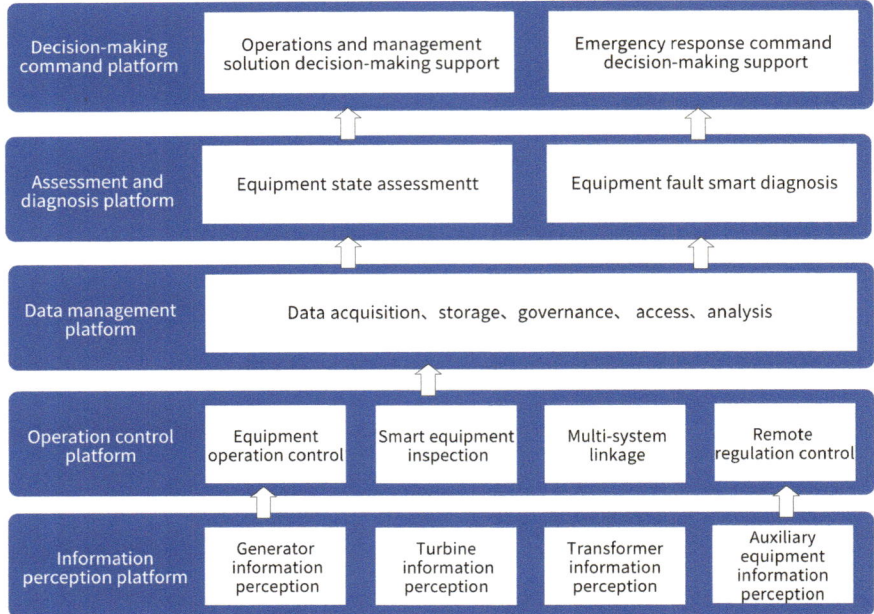

Fig. 6.1 System architecture for Dadu River intelligent equipment operation and inspection

(3) The *data management platform* employs cloud computing technology to achieve data acquisition, storage, governance, and data access services, and through big-data analysis services, to provide data support for intelligent equipment operation and inspection.

(4) The a*ssessment and diagnosis platform* achieves smart assessment of equipment status and smart diagnosis of equipment faults through a fault knowledge base, equipment mechanism modeling, big-data technology, and artificial intelligence learning methods.

(5) The *decision-making command platform* automatically generates equipment operation, maintenance, and servicing solutions, and achieves on-line solution projection through human-computer interaction to provide scientific and efficient decision-making support for equipment management, as well as command and support for on-site emergency handling, through establishment of an expert knowledge base for equipment management and utilization of knowledge reasoning technology.

At the same time, the watershed hydropower intelligent business model for equipment operation and inspection has planning for four major business centers, with the operation management model shown in Fig. 6.2.

(1) *Equipment monitoring analysis center*: Achieves dynamic assessment of equipment health status and timely discovery of equipment anomalies by monitoring

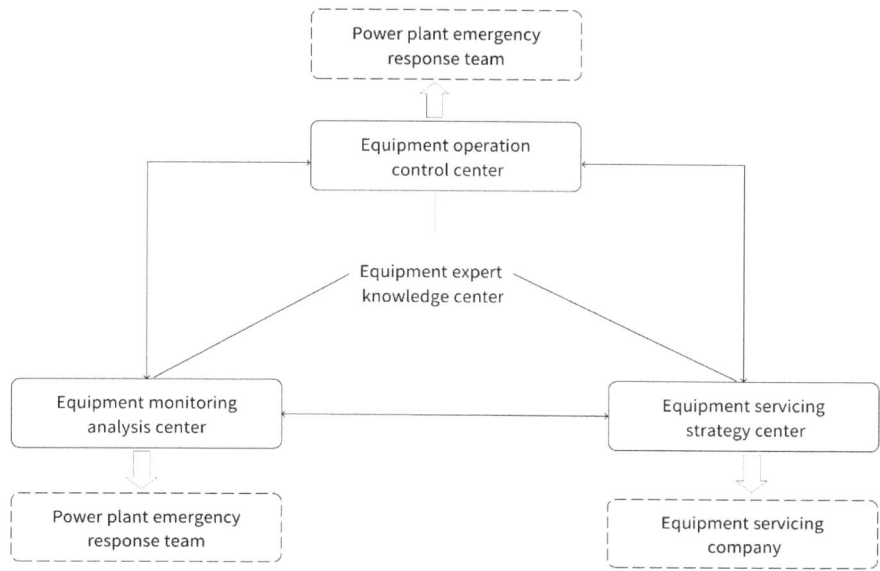

Fig. 6.2 Intelligent business model for equipment operation and inspection on the Dadu River watershed

equipment data in real-time, perceiving equipment operation status, and establishing status analysis models. Based on based on equipment health status forecast and warning models, guides power plant emergency response teams to handle equipment anomalies in a timely and scientific manner, achieving condition-based servicing.

(2) *Equipment operation control center*: Responsible for equipment remote operation monitoring and control, utilizing smart inspection robots and other smart inspection means. With rapid grasp of equipment operation circumstances, under normal conditions this center can control equipment remotely as needed, but can also hand off major operations or equipment anomalies to power plant emergency response teams for timely handling.

(3) *Equipment servicing strategy center*: Capable of pinpointing equipment faults, utilizes the servicing expert knowledge base and fully considers the causes of equipment fault and the degree of damage, automatically generating servicing solutions, while continuously optimizing the solutions through human-computer interaction, to formulate scientific, feasible, and optimized servicing solutions, providing solution support for equipment servicing.

(4) *Equipment expert knowledge center*: Builds and manages expert knowledge bases for equipment operation principles, fault cases, servicing solutions and processes, post- servicing assessments, etc., and unifies the management of equipment-related expert knowledge, while also providing decision-making support for equipment operation, assessment, and servicing.

6.1.2   Build Goals

During the process of building hydropower station intelligent operation and inspection, the Dadu River Company formulated specific build goals for power station smart equipment inspection, business cooperation linkages, smart risk identification, and smart fault diagnosis.

6.1.2.1   Smart Equipment Inspection

Hydropower station inspection is an important means to ensure the safe operation of equipment, and is an effective measure to grasp changes in equipment operating status and the surrounding environment, and to discover equipment defects and critical safety hazards. At the same time, it is also an important pathway to obtain data representative of equipment operating status. With the various equipment of the watershed, situational understanding, and full-dimensional data at the forefront, the forecast and warning of equipment defects and comprehensive perception of unsafe behavior of personnel and environmental state changes has been achieved, by utilizing smart robots, high-definition cameras, smart sensing, and other equipment to simulate manual inspection, and through building smart identification models for image, audio, and other data.

Smart perceptive equipment such as smart inspection robots, high-definition cameras, and wide-band acoustic sensors are uniformly activated and managed through smart inspection system. Real-time perception data is returned for high-definition video, images, sound, temperature, and other perception data of important equipment at key sites, achieving real-time bidirectional transmission of video, pictures, voice, and data.

Smart analysis and processing of the data returned from the devices is carried out utilizing artificial intelligence technology. Through neural networks, machine learning, and other advanced technologies, recognition, processing, and deep learning are carried out for sound, image, video, and other data, along with overall analysis and judgment to determine equipment anomalies and automatically send alert information.

6.1.2.2   Business Cooperation Linkages

By breaking the technical barriers of traditional management in which major production systems are relatively independent, all production system resources of the whole station are integrated. Multi-system linked interaction functions are established, creating smart linkages between core systems such as computer monitoring, ventilation control, fire control, industrial TV, access control, security, and production management. Potential logical connections between systems are developed, breaking the boundaries of each system, and realizing data sharing and centralization, as well as linkage of system functions. Powerful and efficient cross-system linkage functionality is established, fulfilling unmanned on-site operation of generator set equipment under normal working conditions.

At the same time, through site-deployed multi-dimensional equipment status perception and health assessment early warning systems, linked with multiple systems including the smart inspection system, emergency handling of on-site anomalies is achieved, also providing richer information support for remote consultation.

6.1.2.3 Smart Risk Identification

Through a variety of smart wearable devices, linked with sub-systems for daily safety management, threat troubleshooting and management, outsourced project management, dynamic risk early warning, and smart identification of violation of regulations in each power plant, and with an embedded JSA model, the best matching of all elements of "man, technology, environment, and management" is achieved throughout the whole process of production activity, creating an "automated risk identification, and smart risk control" smart safety management model.

6.1.2.4 Smart Fault Diagnosis

Through fault mechanism analysis or big-data modeling, a fault diagnosis model is established to achieve equipment fault diagnosis. Additionally, through establishment of a servicing expert knowledge base, and the building of a servicing decision-making reasoning model, the causes of equipment failure are analyzed and differentiated, servicing solutions are formulated, and the allocation of servicing resources is optimized.

6.2 Key Technology

6.2.1 Smart Inspection of Power Plants Based on Vision and Human-Computer Cooperation

With the development of artificial intelligence technology, application of robotics, computer vision, big-data analysis, and other artificial intelligence technologies is becoming increasingly common and diverse. Currently, more and more industries are looking at application of robotics, computer vision, etc., to replace traditional manual testing, especially for those things that are more difficult to check manually or result in inconsistent judgments. For example, in biomedical engineering, artificial intelligence is widely used in CT imaging technology, processing and analysis of medical microscopic imagery, X-ray images, and other medical diagnosis. In industry and engineering, artificial intelligence is applied in inspection of part quality in automated assembly lines, detection of imperfections on printed circuit boards, stress analysis in elastic mechanics, resistance and lift analysis in fluid mechanics, and identification of workpieces and objects in toxic or radioactive environments. In terms of public safety, artificial intelligence is used in the interpretation of criminal case image evidence, facial recognition, and traffic monitoring or accident analysis.

Taking account of the actual management circumstances for hydropower station equipment operation inspection, the Dadu River Company makes full use of existing robotic automated control technology, smart perception technology, and human-computer interaction technology, combined with advanced technologies such as cloud computing, big-data, IoT, and artificial intelligence, to build a set of smart inspection methods, with deep integration of data modeling and inspection work, a high degree of smartness, and less susceptibility to human interference, in order to provide the necessary technical guidance for manual analysis and decision-making, ensuring the safe and stable operation of hydropower station generator set equipment.

6.2.1.1 Technological Characteristics

By adopting a method of multi-source perception that inter-combines "mobile perception + fixed monitoring" of hydropower station equipment status and environmental information, smart inspection technology replicates the four diagnostic methods of "observe," "listen," "inquire," and "touch" in the daily equipment inspection work of operations and maintenance personnel, providing the smart inspection system with functions such as sight, hearing, smell, etc. Based on computer vision, video processing and recognition, acoustic signal analysis, and other smart technology, a system "brain" is developed—a smart inspection control platform, carrying out recognition, processing, and deep learning of the captured images, video, sound, etc., accurately diagnosing anomalous equipment conditions, forming algorithmic models for image and video stream recognition applicable to hydropower scenarios, so that the platform has the "smart" ability and is thus able to replace manual inspection work by personnel. The primary characteristics are:

(1) *Brighter "eyes"*: With multiple high-definition cameras as the eyes of the smart inspection system, the system can identify leaks or dripping of various meters or in various environments.
(2) *Sharper "ears"*: Wide-band acoustic sensors perceive abnormal sound generated in the special environment, and daily data accumulation and model training gives the system the ability to identify anomalous sounds.
(3) *A more sensitive "nose"*: Various smart sensors allow the system to perceive the odors of specific gases.
(4) *More acute perception*: High-sensitivity infrared temperature measurement technology allows the system to be more perceptive of abnormal equipment temperature increases.

(5) *More specialized "diagnosis"*: Daily multiple data comparison and analysis in employed to identify equipment anomaly categories and degree of anomaly, molding the smart inspection system's diagnosis ability.

6.2.1.2 Multi-dimensional Smart Acquisition Technology for Inspection Data

Given the hydropower plant building's cavernous structure and the complexity of the plant's vertical layers and numerous horizontal rooms, as well as the generator and auxiliary machinery, along with oil, water, and gas equipment, all with their criss-crossing and irregular layouts, the Dadu River Company employs an inspection robot (as shown in Fig. 6.3), with visible light camera, infrared thermal imaging, sound acquisition, gas sensors, and other inspection devices, to create multi-dimensional data smart acquisition technology to achieve effective fusion and integration of video, audio, temperature, vibration, and other data, and the integration of complete equipment operation figures within the hydropower station data platform. The smart inspection robot is critical equipment for inspection work, and also critical equipment for safety supervision of the power plant site environment and equipment, and is a comprehensive control system to realize on-site unmanned inspection and automatic anomaly warning, which can significantly reduce intensity of labor for operations personnel.

The front-end acquisition instrument of the robot has perception that emulates biological perception functions such as hearing, seeing, smelling, and touch, to achieve the perception of elements such as the field-of-view, sound, and environment (oil leakage, water leakage, foreign items, and other anomalous states) of equipment in the area.

Other smart inspection devices include important supplementary system monitoring units composed of smart components such as a thermal imaging cameras, wide-band acoustic sensors, and temperature and humidity sensors. These are mainly applied to ventilation covers, ventilation ducts, lower bearing brackets, turbine chambers and other restricted spaces, with the capability to sense temperature, abnormal sounds, pooling water, oil leaks, vibration, and other environmental conditions. Hydropower station inspection is achieved through multi-dimensional smart acquisition technology, with station-wide full detection coverage and no overlooked dead spots, meeting requirements for full-element, full-time, and full-dimensional smart perception of equipment health status.

Currently, smart inspection robots used in hydropower plants in the Dadu River watershed are divided into two primary types: tracked robots, and wheeled robots. Wheeled robots are mainly used in the main plant building spaces that have wider and more even ground, such as at the generator and turbine levels of the power plant. Tracked robots are mainly used in narrow spaces such as cable galleries and ventilation ducts.

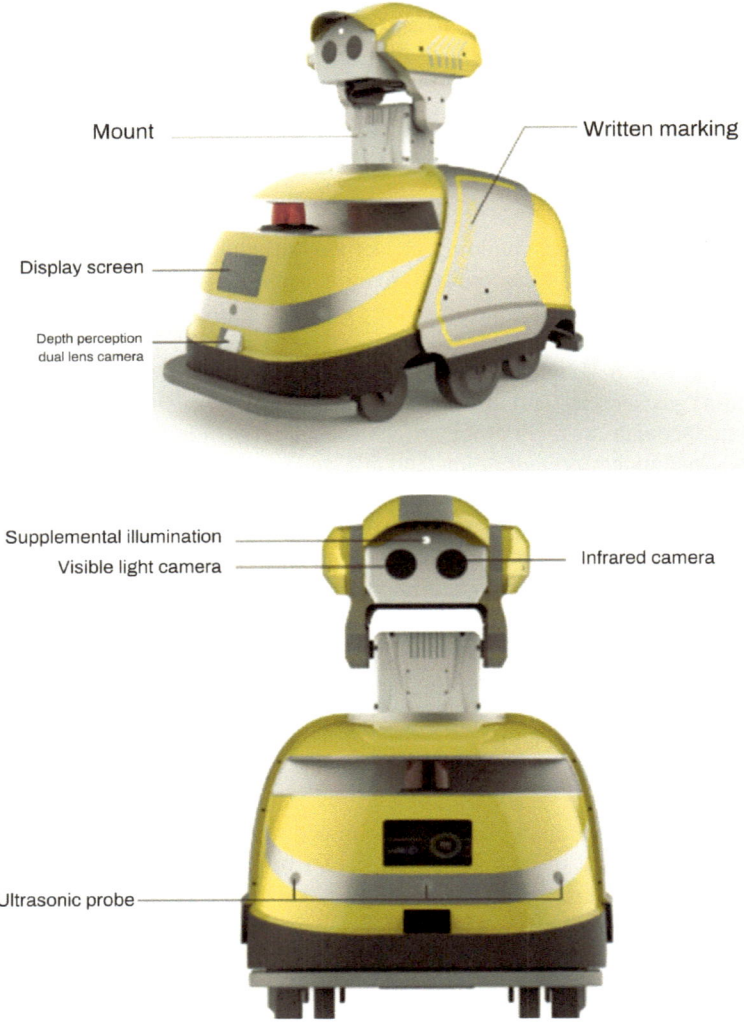

Fig. 6.3 Hydropower station smart inspection robot (wheeled)

The self-developed smart inspection robot is mainly composed of a head, obstacle avoidance device, charging device, travel mechanism, etc., with the following basic functions:

(1) *Smart video image recognition*: The inspection robot is equipped with a high-definition visible light camera, and uses computer vision technology to capture images and videos and to carry out recognition analysis of mechanical meters, digital display meters, liquid level meters, circuit breakers, and valve positions, etc., achieving smart recognition.

(2) *Precise infrared temperature measurement*: The inspection robot is equipped with an infrared imager to carry out acquisition, smart recognition, and analysis of infrared images of equipment and key sites needing measurement, achieving temperature monitoring, alert, and trend analysis for the relevant equipment.

(3) *Smart audio recognition*: The inspection robot is equipped with directional audio pickups. Based on the direction of the sound source, the target signal is picked up from the midst of mixed signals, picking up only the sound signal propagated in a specific direction and shielding against noise and interference from other directions, to achieve the effect of enhancing the target sound. Based on the propagation characteristics of sound waves, using the temporal, spatial, and frequency characteristics of the sound signal, along with research on directional audio pickup technology, real-time analysis is carried out in the smart backend to discover anomalous states.

(4) *Two-way voice communication*: The inspection robot is equipped with a two-way voice system and equipped with emergency loudspeakers and listening microphones for two-way intercom between the system backend and field personnel, realizing remote site monitoring and command.

(5) *Autonomous navigation and positioning*: Laser navigation and SLAM algorithmic positioning are used to achieve autonomous robot navigation and improve positioning accuracy.

(6) *Automated charging*: When the power level in the inspection robot gets low, it can automatically return to its charging station to recharge.

(7) *Early warning and alert*: By analyzing historical data statistics and trends, the inspection robot can predict potential equipment faults and deficiencies in the operating environment in advance, and issue early warning information in a timely manner.

The inspection robot also carries gas-sensitive sensors or sensor arrays to convert gas and concentration information into electrical signals, and utilizes pattern recognition methods to analyze and determine odor category. The relationships between odor and equipment is determined through the system's on-line learning and analytics technology, accomplishing monitoring and identification of anomalous states of on-site equipment.

6.2.1.3 Smart Image recognition, and Mechanism Modelling for Smart Allocation of Inspection Tasks, in a Real-World Environment

In view of the performance characteristics and development circumstances of the Dadu River hydropower station equipment operation status situational understanding system, and directed at full-dimensional data related to hydropower station equipment operation status monitoring elements and environmental characteristics as research objects, a smart perception system based on perception elements like robots and auxiliary inspection equipment was developed, which can smartly identify and logically correlate front-end perceived information, emulate the thinking modes of the human brain, achieve autonomous analysis, autonomous learning, logical induction, correlation analysis, and can carry out linkage with other systems, to realize inter-analysis and interoperability between systems. At the same time, the internal thinking module employs standardized design, so as to be conducive to replication and promotion.

Based on the ordering and analysis of the application functions of each smart device and system within the hydropower station inspection task scenario, a work model for smart task assignment under each typical scenario of hydropower station inspection operation was built. Through cooperative work between people and smart devices and systems, the efficiency of personnel operations is effectively improved, the collaboration barriers between on-site field personnel, backend managers, experts, and smart devices are broken, and a unitary system of operation cooperation is achieved, covering the whole cooperative process of smart acquisition and identification of hydropower station site operation data, active risk early warning, operation information assisted decision-making, closed-loop management, and structured data storage, etc. At the same time, using IoT and network communication technology, and while ensuring data security, on-site operation data is comprehensively stored, aiding safety and quality control. Inspection operation technicians, backend managers, and intelligent analysis terminals are provided with applications for real-time team collaboration, on-site command, expert guidance, safety supervision, interactive decision-making, and risk warning, so as to effectively improve on-site work efficiency, guide safe operation, reduce human errors, with traceability for accident causes, and providing a guidance aide for inspection operations.

6.2.2 Power Station Work Safety Risk Process Control Technology Based on Smart Wearables

Safety is a fundamental prerequisite for the establishment and development of power generation enterprises. In the hydropower station, safety management refers to the manager's control actions directed towards all relevant elements in order to ensure the safety of the hydropower station, particularly dynamic control. In accordance with current productivity levels, people are one of the most important drivers of productivity in production practice, so personal safety of personnel is the first priority of safety management.

Personal safety, equipment safety, and environmental safety are complementary and mutually reinforcing. Operations of humans act directly upon equipment, and the process and results of the operation will produce changes in the environment. The state of a piece of equipment may change at any time, and the same equipment that is not charged with electricity one second, may the next second become an electrified killer. The state of the environment will also change with time and space, a few days before a specific location may have been flat ground, and a few days later that location may be a deep pit. Various chaotic environmental safety factors can unknowingly undergo any number of transformations. How to have people work safely in intricate and unsafe environments is an important proposition for safety control.

The Dadu River Company has built a safety risk control center based on multi-dimensional information fusion perception, which changes the traditional safety management model by identifying the risks or risk factors that exist in production operation, qualitatively and quantitatively analyzing their severity, formulating and implementing a series of measures and regulations to manage production and operation activities, and effectively controlling unsafe factors before they occur.

First, the workflow is simple and transparent. In traditional operations, safety work approval is usually by layer upon layer of paper documents passed around. Actual implementation of safety work is usually through item-by-item on-site review and confirmation by dedicated safety management personnel. This type of management is generally not very efficient. The new approach employs mobile internet technology. Reviewers for different constituent elements can conveniently and efficiently carry out processing, and information changes are updated transparently in real-time. Through smart analysis of the status of rectification work via smart image comparison technology, the safety management process can be back traced, and work traces are not easily tampered with, improving the efficiency of safety hazard dissemination and rectification, and also making the entire work chain more transparent and traceable.

Secondly, the paper logbook is completely eliminated. In traditional management, management personnel on-site often use hand-written logs and photographic evidence, but paper logbooks are easily damaged or lost at the production site, and are not convenient to carry. In the new management approach, personnel for each management constituent element can record logbook information in real-time through a mobile terminal, improving on the inconvenience of traditional management in a real way.

Additionally, automated early warning averts omissions. The interconnectivity between the specialized data centers means that sources of safety management data are more complete and diverse, and the information basis for early warning judgements is more robust and reliable. By linking with sub-systems for daily safety management, threat troubleshooting and management, outsourced project management, dynamic risk early warning, and smart identification of violation of regulations in each governed power plant, and interacting dynamically with real-time data from various smart equipment such as such as industrial television systems, standard image libraries, smart safety helmets, smart safety belts, smart safety ladders, and smart bracelets (shown in Figs. 6.4 and 6.5), the best matching of all elements of "man, technology, environment, and management" is achieved throughout the whole process of production activity, creating an "automated risk identification, and smart risk control" smart safety management model.

The Dadu River Company safety risk control system is divided into three usage levels: Company headquarters, base-level units, and individual employees.

The company headquarters primarily utilizes the hardware network system, application support system, and information safety system, for comprehensive collection,

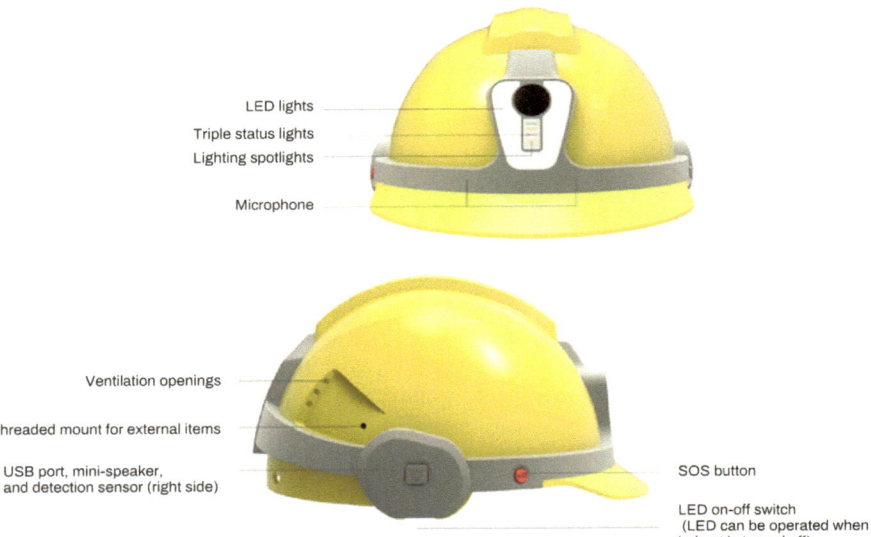

Fig. 6.4 Smart safety helmet

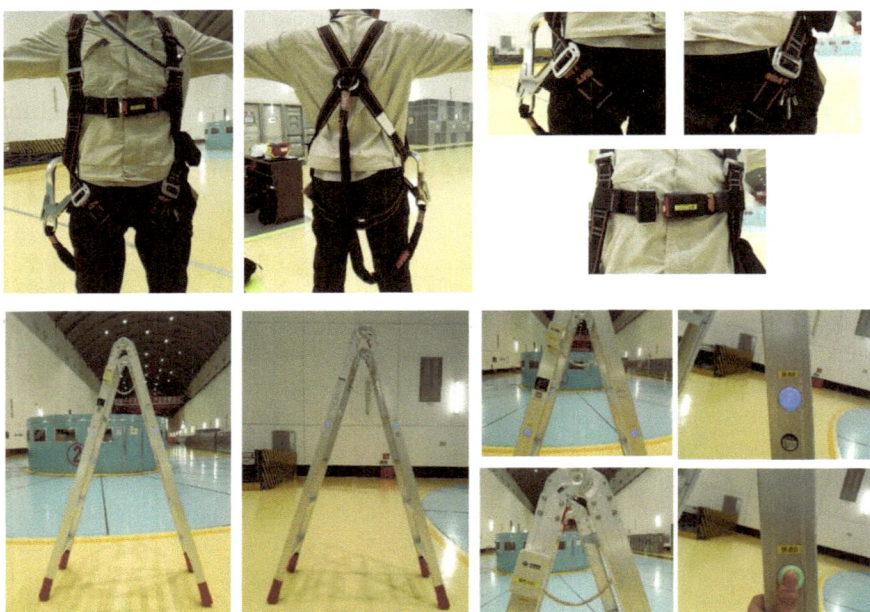

Fig. 6.5 Smart safety belt and smart safety ladder

analysis, and mining of various types of relevant safety risk data of the company head-quarters and base-level units, to achieve safety risk control and early warning, and the formation of four safety management platforms: The base-level business platform, core business platform, decision support platform, and mobile application platform. Base-level unit safety management automation of early warning alerts, electronifi-cation of information processing, standardization of threat troubleshooting, mobi-lization of safety inspection, clarification of job responsibilities, and quantification of performance assessment, are achieved primarily through data-driven approaches. This effectively strengthens the departments and teams to rigorously take on prin-cipal responsibility for safe production, solves problems at the base level, eliminating threats while still in an initial state, and uses information technology to drive the genuine adoption of safe production threat troubleshooting.

Employees' personal and office PCs share system production data, using informa-tion technology and mobile terminals to improve the effectiveness of safety managers at all levels, and reduce the burden for base-level personnel. This is closely integrated with safety-focused operations, designed and implemented in terms of data-driven uniformity, cooperability, convenience, ease of use, and sense of experience. Through mobile applications, on-site briefing and pre-operational risk analysis and pre-control can be achieved. Timely work alerts and notices can be pushed out. Mobile review and approval shortens process approval time. Rapid upward reporting of threat trou-bleshooting and site review can be made, improving efficiency of on-site work, and the status of safety risks can be easily queried at any time or location. A dynamic risk evaluation model based on a semi-quantitative assessment is able, in real-time, to reflect value-at-risk for the power plant or for each production area of the power plant. This risk dynamic assessment model achieves the functionality of automated graded matching of hazard sources (points).

Through the categorization of hazard sources (points) and the coding of safety duties of personnel at all levels, identified risks are automatically matched to the specific responsible person for the four hazard source control levels, making the safety risk graded control model smart and efficient, achieving automated risk warning, accomplishing whole-process management of operational safety risks, from assess-ment to control, and smartly calculating the complete area risks. Real-time dynamic early warning provides substantiation for managers to control safety risks.

6.2.3   Equipment Health Status and Fault Analysis Warning Technology Based on Multi-dimensional Information Fusion Perception

In the process of hydroelectric power plant equipment operation and inspection, a large quantity of data will be generated, such as equipment servicing records, and fault cases. This information not only effectively promotes enhancement of the servicing department's servicing work, but can also provide reference information for decision makers, and raise the efficiency and accuracy of maintenance assurance.

Various monitoring and control data, equipment status data, production automation data, and safe business management data in Dadu River watershed provide powerful data support for safe production management. Yet with the acceleration of the informatization process, the multiple sources, heterogeneity, and vast quantities of data also bring difficulties for efficient data integration and sharing, and the relevant business personnel cannot obtain useful knowledge from the data quickly and effectively, and thus cannot carry out more accurate and efficient servicing work based on the data. A knowledge graph provides a means to extract structured knowledge from among massive data and enables machines to have cognitive ability, thus making knowledge graphs one of the key technologies for data analysis. The introduction of knowledge mapping technology into the field of hydropower plant servicing in the Dadu River watershed solves the current difficult issues of efficiently extracting key knowledge from massive data in the servicing data center and equipping machines with cognitive ability.

The Dadu River watershed knowledge map is composed of key knowledge interconnected in triples, either "entity-relationship-entity" or "entity-attribute-attribute value." In this, entities are the most basic elements, which are connected two-by-two in relationships, thus forming a structured knowledge network. In production operation, equipment, such as turbines, excitation wheels, etc., are entities in the knowledge graph. An entity can contain multiple attributes, which describe the possible attributes, characteristics, properties, and parameters of the entity, such as commissioning time, installation location, manufacturer, etc. Each entity has a unique ID to distinguish it from other entities. Each attribute value corresponds to the intrinsic characteristics of the entity, and the relationship is used to describe the association between entities.

Take the equipment knowledge graph of a transformer as an example (as shown in Fig. 6.6). The equipment major defect A0001 is an entity, the servicing person is an entity, and the triple *major defect A0001—servicing needed—servicing person* is an example of an "entity-relationship-entity" triple. Also, the rotor is an entity, the manufacturer is an attribute, and the specific manufacturer is an attribute value. The triple *bushing-manufacturer-specific manufacturer* constitutes an example of an "entity-attribute-attribute value" triple.

Based on the servicing data center's data sources, a combination of both top-down and bottom-up approaches is employed to build the knowledge graph for hydropower equipment servicing. In other words, the data schema is first defined manually, and then information is extracted for the specific relationship of the data schema. Servicing data center data sources are given precedence and are used first, and then these are supplemented by other general data sources. Construction of the knowledge graph occurs from multiple data sources, and key knowledge in the data is extracted through the three stages of information extraction, knowledge fusion, and knowledge processing, and then deposited into the data and schema layers. In order to ensure the knowledge graph's real-time character, it is also necessary to iteratively employ the full technical process to update the assembled knowledge graph.

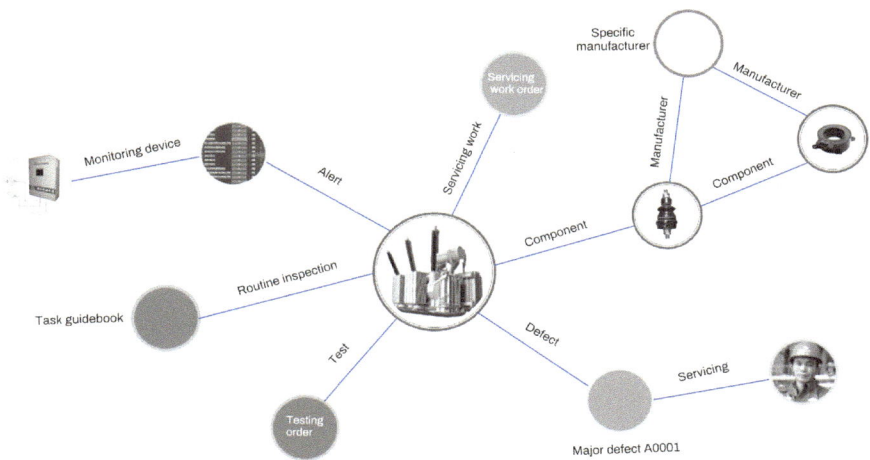

Fig. 6.6 Knowledge graph for transformer equipment

Knowledge graphing of hydropower equipment servicing provides an effective means to utilize and manage vast heterogeneous data sources, and a large amount of equipment servicing data becomes well expressed and organized. Applying this to the field of equipment servicing, a feature store for autonomous learning of hydropower generator set faults was established, and a real-time equipment health degree assessment model and an approached for early warning of equipment health trends were developed.

Equipment Data Analysis: The Dadu River Company first comprehensively analyzes data from a single item of equipment. Collection and analysis of multifaceted status data for the item are carried out over a period of time, and a summary is made of equipment operation patterns. Based on this, a trend analysis is performed across an aggregate of generator sets at the same measurement point. Since equipment models, layouts, and configurations of the company's main hydropower stations are essentially similar, selecting data from multiple sample groups for identical or similar equipment for analysis and pattern summary has greater validity, and is more helpful for equipment analysis and summary, and for providing guidance for operational maintenance. With a pattern for equipment operation, reasonable dynamic values for equipment can be formulated. Under different working conditions, dynamic values can be fixed by summarizing a combination of water head, carried load, ambient temperature, seasonal changes, and other factors. Decision-making for condition-based servicing is provided through analysis of equipment data, judgement of equipment health condition, and assessment of equipment state.

Establishment of Fault Library: Hydropower station equipment historical data, operation data, servicing data, and equipment fault data are integrated and collected to establish an equipment fault library. The fault library widely accumulates fault data and provides samples for research into fault diagnosis methods. The fault library shows the operation status and data for both before, and during, the equipment fault, and records the process of analysis, methodology, direction, and fault handling measures. Through the fault library, specialists from different power plants interact across space to achieve rapid analysis and resolution of equipment faults.

Fault Early Warning: With the equipment fault library, combined with typical equipment faults occurring in and outside the watershed, collected key characteristic data of the equipment before, during, and after the fault are analyzed, and characteristic parameter change patterns are summarized to build the fault early warning model. When characteristic parameters of the equipment change in ways that correspond with signs of coming failure, a fault early warning is issued so that operations and maintenance personnel can take timely measures to eliminate the equipment fault at an initial state, and prevent the equipment fault from escalating. At the same time, the system suggests next steps or changes in direction, so as to provide decision-making support for operations and maintenance personnel regarding measures to adopt.

State Quantification: Quantitative equipment state assessment technology uses a partitioning assessment method, which is a technique for carrying out appraisement of a set of characteristic quantities reflecting the health status of the equipment. The conclusion of the assessment is expressed in four grades: A (good), B (qualified), C (anomalous), and D (dangerous). Each characteristic quantity corresponds to a specific servicing content, allowing direct pinpointing of faults and defects. This kind of technology is an algorithmic strategy for state assessment that is well suited for implementation on computer systems, especially for on-line execution on platform systems, to form fully automated on-line fault diagnosis systems capable of engineered configuration.

6.2.4   Equipment Health Status Decision-Making Support Technology

Traditional hydropower servicing employs a planned servicing model. During management, each work type acts as a boundary to divide the different specialized teams. At servicing time, each team individually completes the various specialized sub-projects. For each service project, management authority for the drawing of personnel, work arrangements, and reward distribution all rests within each team, and the management of the teams is relatively independent. Under the pressure of diverse and intricate service projects, shortage of human resources, and market competition, the traditional service model cannot adapt to the current rapid development of hydropower equipment service. For this reason, the Dadu River Company has explored a new service model and management approach.

Innovative management model: Centered around the Dadu River Company's hydropower equipment condition-based servicing approach, through independent innovation, a set of "multi-item unitary integration" intelligent servicing decision-making and execution control models has been established. Here, "unitary" refers to the control model as a whole being based on same equipment operation and servicing workflows. Process critical points, servicing decision-making strategy, and execution control requirements are uniformly defined for each workflow, in whole forming a unitary set of integrated work association trees, from decision-making command, to execution control. "Multi-item" refers to the specific work elements in this work association tree, such as human resources, financial resources, equipment and materials, project management, standards, safety supervision, risk control, and other aspects.

Innovative application of technology: Adoption of the "multi-item unitary integration" intelligent servicing decision-making and execution control models employed many innovative technologies.

At the data layer, equipment health status figures were provided through real-time equipment health status monitoring, diagnosis and analysis, and trend early warning technology, while at the same time, cross-specialty and cross-departmental core enterprise resource figures such as human resources, finance, materials, equipment, and projects, were brought together.

At the application layer, centered around the need for whole process management decision-making support and execution control for hydropower equipment operation and inspection, decision libraries for both equipment fault and for equipment health trend risk were established based on current and possible future equipment health status risks. Additionally, decision libraries were established for emergency repair execution control, planned servicing execution control, and normal operational maintenance, according to typed and graded equipment health status risks.

In terms of human-computer cooperation, centered around the critical process points of each workflow, decision-making analysis reports can be automatically and smartly sent to the corresponding positions and personnel. Visual analysis and decision-making statistical reports can be provided for relevant managers on a regular and topical basis, along with periodic or dynamic equipment condition-based servicing strategy reports for each generator set in the hydropower stations.

Decision-making support sustainability assurance: The implementation of a system for servicing intelligent decision-making and execution control did not happen in one easy step. Rather, it required continuing work in the collection and aggregation of data figures related to "multi-item unitary integration" and optimization and improvements to indicator models.

For the needs of intelligent decision-making and execution control of equipment operation and inspection, on the one hand, the Dadu River Company built a general technical rule library for data, formulating general technical rules for data governance in terms of valid fields, principally including content such as data uniqueness, integrity, consistency, timeliness, etc., while on the other hand, we built a "multi-item unitary integration" work data quality rule library to verify data validity, accuracy, and the precision of data profiling (verification of validity of maximum value, minimum value, average value, summary value, etc.). Finally, combined with the general technical rule library and work data quality rule library, the company built automated and intelligent data verification tools, providing regular periodic data governance reports. After servicing, a combination of computer system and expert experience together scores the servicing process and continuously improves the decision-making system.

6.3 Case Application

6.3.1 Successful Identification of Discharge Phenomenon in the High-Voltage Station Service Transformer in Generator Set 5 of Pubugou Hydropower Station

6.3.1.1 Overview of the Smart Inspection System at Pubugou Hydropower Station

Equipment inspection is an important work of hydropower station operation and management. Traditionally, the hydropower station employs manual inspection rounds. That is, inspection personnel need to carry out periodic checks of the equipment and locations requiring management and monitoring. Inspection personnel are required to go to the designated location to check whether the equipment is normal, whether there is an odd smell or noise, whether the environment is safe, etc. Inspection personnel make judgement based on past experience, and either promptly handle the discovered deficiencies or report them to relevant technical personnel. This traditional inspection method is subject to interference from the external environment, and the high and ultra-high pressure environment of the hydropower station poses a relatively large threat to the personal safety of the inspection personnel. At the same time, inspection results will be affected by the staff's operational capacity, work experience, mental state, and other factors. Inspection errors and omissions do occur at times, causing significant economic losses, affecting the safety and stability of the entire hydropower station. Additionally, manual inspection is characterized by great labor intensity, poor inspection quality, high management costs, slow emergency handling, and other shortcomings. Therefore, manual inspection as an operation means can no longer guarantee the safe and stable running of the power generation system.

In response to this situation, Pubugou hydropower station widely applies increasingly mature image recognition, artificial intelligence, and other technologies, in the research and development of a hydropower station smart inspection system. This system innovatively brings in smart inspection robots, fixed video cameras, sound capture devices, and other sensing equipment, adopting a "mobile + fixed" deployment mode, to achieve smart acquisition of hydropower station video, audio, infrared, sensing, and other full-dimensional inspection data, to emulate the four diagnostic methods of "observe," "listen," "inquire," and "touch" of the daily equipment inspection work of power personnel. By acquiring multi-dimensional information of hydropower stations such as image data, infrared thermal imaging data, sound data, temperature data, etc., and employing computer vision technology and machine self-learning technology, an artificial intelligence smart scheduling engine has been developed for high volume data and tasks, forming a mechanism model for audio and video smart identification and smart allocation of inspection tasks for the complex environments of hydropower stations, making it "smart," thus gradually replacing manual inspection work.

The smart inspection robot is the primary device for implementing equipment and site environment safety supervision, and the smart inspection system is an overall control system to achieve power plant unmanned inspection and automatic early warning, which further reduces the labor intensity for operation personnel. The front-end acquisition instrument of the robot has perception that emulates biological perception functions such as hearing, seeing, smelling, and touch, to achieve the perception of elements such as the video, audio, and environment of equipment in the area, and identify anomalies through artificial intelligence technology. For circumstances where the robot cannot go into special restricted spaces in the hydropower station, "electronic eyes" such as industrial TV systems, sound sensors, and infrared thermal imaging equipment are widely applied to achieve smart perception of temperature, unusual sound, water pooling, oil leakage, and vibration in restricted spaces such as turbine-generator ventilation covers, ventilation ducts, lower bearing brackets, and turbine chambers.

6.3.1.2 Basic Case Background

At 8:55 on January 4, 2020, the upper generator set reported "5B main transformer No. 1 protection low-voltage side zero-sequence voltage alert," "5B main transformer No. 1 protection device alert." The whole station AVC function was withdrawn. The fault oscillograph showed that the C-phase voltage of No. 5 high-voltage station service transformer was zero, and on-site sampling of low-voltage side C-phase voltage of the main transformer 5B protection device was zero. At the same time, the smart inspection system industrial television front-end camera discovered that Pubugou hydropower station No. 5's high-voltage station service transformer 5CB internally evidenced obvious discharge phenomenon, and the system issued graded alert information to relevant operations and maintenance personnel and technical management personnel. At 9:05, equipment management center personnel arrived on site to guide emergency handling work. At 9:08, production command center control instructions were received: Take 5B out of service. Through a comprehensive inspection of the transformer body, it was found that the 5CB high-voltage side C-phase to IPB connection cable had overheated and showed serious burning. The external insulation on one of the connection cables had melted, and the 20 kV connection cable copper conductor was bare, and the epoxy plate near the connection cable showed charring, as shown in Fig. 6.7. At 21:24, Dadu River Servicing Company completed replacement of 5CB's C-phase burned cable and epoxy plate, and had added head-shrink casing over the three-phase connection cable to increase

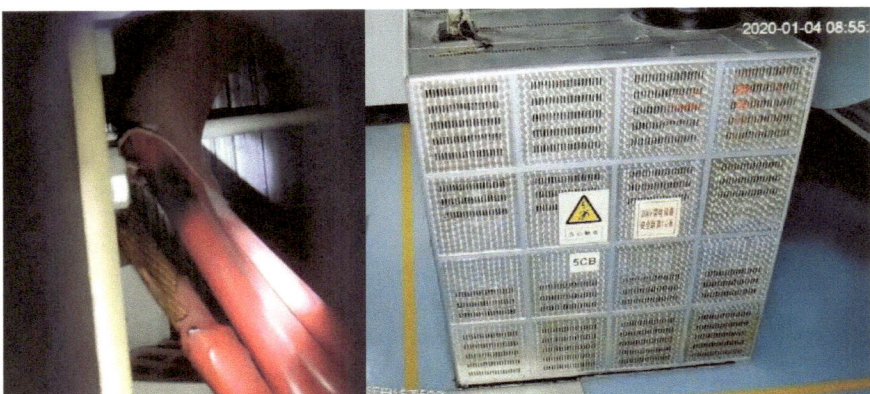

Fig. 6.7 Transformer discharge fault images

insulation. Both high- and low-voltage sides were reconnected, and it was ready to supply power.

6.3.1.3 Application Results

The hydropower station smart inspection system employs a "smart inspection robot + fixed video cameras" mode to gradually replace the traditional manual inspection. The system makes up for insufficiencies in hydropower station equipment self-collected data by acquiring information from the system itself and from its surrounding area. The smart recognition framework, built with deep learning as the core, achieves smart recognition of 10 types of characteristic quantities, such as the "three leaks," status indicators, sound, etc., improving the level of smart perception of equipment health status, realizing real-time perception and early warning of field equipment defects and abnormal conditions, and improving the timeliness of defect discovery, preventing incident escalation and equipment damage caused by late discovery and handling, avoiding unplanned accident downtime, and simultaneously avoiding large economic losses.

6.3.2 Discovery of Signs of Anomaly on Gongzui Power Station Fixed Guide Vane Turbine Four Months In Advance

6.3.2.1 Intelligent Servicing System Overview

Dadu River Company built an intelligent servicing system based on self-developed Hypersphere modelling, primarily composed of health assessment, big-data trend warning, and other functions. The Hypersphere modelling algorithm is a machine learning algorithm that uses all relevant measurement points for hydropower plant main equipment, combined with data from different historical operating conditions, to establish various models, and automatically carry out on-line evaluation of the real-time status of industrial objects, synthesizing real-time status of turbine-generator sets into a 0–100% assessment value, called "health level HPI." The system analyzes the historical data and builds a hypersphere model to obtain a health baseline value "Hth," which is the standard for assessing the operational health status of the turbine-generator set. During operation, when the HPI health level of the equipment deviates from historical safe operating conditions, the system issues early warning of potential fault before the alerts for individual equipment parameters, and at the same time automatically gives the ranking of associated measurement points that led to the change of generator set status, and carries out automated association analysis of early potential faults.

The intelligent servicing system has been gradually deployed in the power stations of the Dadu River watershed starting in 2014, providing early warnings to many faults, and achieving good results.

6.3.2.2 Basic Case Background

Gongzui hydropower station is located in the middle and lower reaches of the Dadu River at the junction of Shawan District and Ebian County, Leshan City, Sichuan Province, China. The first generator set produced electricity in February 1972, and construction completion and station commissioning was in 1978. With a design head of 48 meters, the station is equipped with seven Francis turbine-generator sets with individual capacities of 110,000 kW and a total installed capacity of 770,000 kW.

With the increasing seriousness of silting of Gongzui reservoir and increase of sediment passing through the generator sets, the abrasion of components of the turbines which pass water has become more and more grave. Frequent recurring problems, such as penetrating cracks in the turbine top covers, sealing rings, and runner blades, and cracking of the water diversion cover plate, have caused enormous economic losses, and seriously threatened the safe, economical, and stable operation of the generators sets.

On June 4, 2016, signs of anomaly were observed when, during operation, the health curve of generator set No. 7 of Gongzui power station suddenly started to

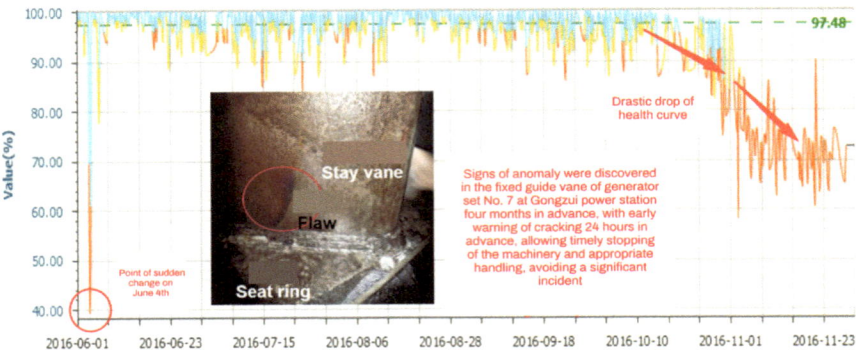

Fig. 6.8 Health level assessment and trend early warning

fall. Being precisely during flood period, a key period for high power generation, according to traditional management methods, given the unclear extent of damage, the set would need to be immediately shut down and serviced. However, the application of intelligent servicing system enhanced Dadu River Company's equipment risk pre-control capability. In this case, after overall consideration of equipment health and power generation benefits, it was judged that immediate shut down for servicing was not required, and through the equipment control center's equipment health assessment, equipment risk control measures were established: While generator health was the range of 90–80%, the site would implement a protective operation strategy, avoiding the load range where abnormal noise occurred (93–110 thousand kW); when the equipment health dropped below 80%, the generator set would immediately be shut down and the issue addressed. Entering November, the health level trend of the generator set matched the prediction, with health dropping rapidly below 80%, as shown in Fig. 6.8. The production command center decisively stopped the set for inspection and found that there were multiple cracks in the fixed guide vane and broken pole key in the rotor pole, and appropriate servicing was arranged.

6.3.2.3   Application Results

Early warning and processing by the intelligent servicing system in this incident ensured that there was no unplanned equipment downtime and effectively ensured equipment and personal safety, bringing direct economic benefits of more than 2 million yuan and indirect economic benefits of more than 17 million yuan.

6.3.3 *Successful Proposal of Servicing Strategy for Stator Failure at Dagangshan Hydropower Station*

6.3.3.1   Application Background

Stator partial discharge monitoring systems have extremely high requirements regarding quality and installation technology for the related sensing equipment, and the stability and accuracy of the monitored generator greatly affect its monitoring results, leading to unsatisfactory application results for partial discharge monitoring systems in hydropower stations.

With 49 generator sets already commissioned or under construction in the Dadu River watershed, involving an installed capacity of about 12 million kW, and considering several cases related to stator insulation failure that had occurred one after another during the previous few years in the power generator sets of the watershed, the need to carry out research on smart monitoring and early warning of stator local discharge became urgent. After detailed research on the application of stator discharge related technology in the hydropower industry, and numerous discussions from various aspects on the feasibility of stator discharge technology, a pilot project for generator set stator partial discharge on-line monitoring technology was carried out to build a stator partial discharge smart monitoring and early warning system to help power plants make objective judgment on the condition of generator set stator insulation, and additionally provide valid substantiation for equipment servicing.

The smart monitoring and early warning system for stator partial discharge for a hydropower turbine-generator set is composed of a single set's discharge monitoring device and a unified partial discharge monitoring workstation. Coupling sensors (as shown in Fig. 6.9) are installed in the stator to monitor for signs of partial discharge, and data communication is carried out through the partial discharge monitoring computer, and specialized analysis is carried out on the partial discharge data and relevant results are displayed at the partial discharge monitoring workstation.

Starting from the most critical but most difficult to solve problem of insulation, the smart monitoring and early warning module for turbine-generator set stator partial discharge achieves real-time monitoring, smart analysis, and accurate prediction of stator partial discharge through reliable technical means under the actual working conditions of generators. This module puts forth targeted servicing suggestions, which greatly raises the intelligence level of equipment fault diagnosis and decision-making support, and improves the equipment management level on the Dadu River watershed. The smart monitoring and early warning interface of the stator partial discharge is shown in Fig. 6.10.

6.3.3.2   Basic Case Background

During Dagangshan generator set operation, the set partial discharge monitoring system was fully utilized to observe and track the partial discharge of generator set No.

Fig. 6.9 Coupling sensor
installation diagram

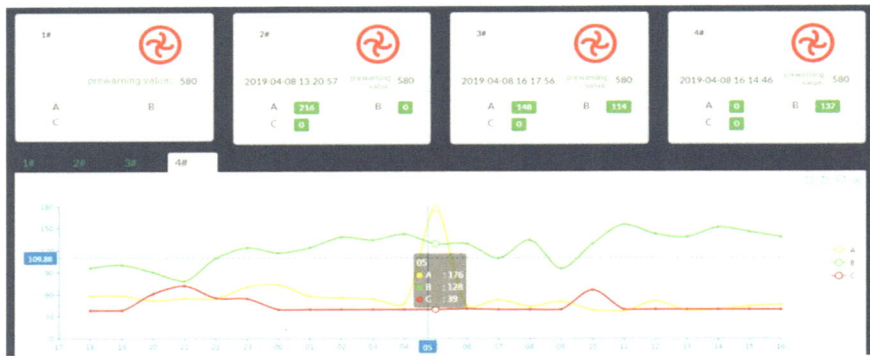

Fig. 6.10 Smart monitoring and early warning of stator partial discharge

1, and the partial discharge data was analyzed at regular periods. The "Dagangshan Generator Set No. 1 On-line Partial Discharge Test Report" showed that the partial discharge of Dagangshan generator set No. 1 had been at a relatively high level since set commissioning, far exceeding the level of 95% (768 mV) of the partial discharge amounts of other same model sets, with the highest partial discharge being close to 4500 mV. The analysis showed that the generator set was experiencing interphase discharge, and that the partial discharge value was high, but stable with no trend towards growth. From this a servicing strategy was composed: Close attention would be paid to the partial discharge development trend, and necessary inspection and handling would happen during the servicing period, with strong attention on traces of interphase discharge, and additionally complemented with offline partial discharge testing applied to the rated line voltage level.

During the servicing period, the stator and rotor were cleaned and painted in accordance with the servicing strategy. At the same time, the insulation of the jumper wire at the end of the stator bar was meticulously checked, and a total of seven places were found to have loose, softened, and delaminated insulation. Insulation material on the faulting component was stripped and new insulation was applied. After servicing and retesting, the partial discharge level returned to normal.

6.3.3.3   Application Results

Smart monitoring and early warning of stator partial discharge can prevent generator failure caused by stator discharge deterioration to the maximum extent, and reduce the human and material resources required to locate and deal with the failure. From the successful case of Dagangshan generator set No. 1, more than 10 days were saved from this single instance of generator set servicing, which brought economic benefits of 39 million RMB.

Chapter 7
Intelligent Operation of Hydraulic Structures

7.1 Overall Approach

Operational management of Dadu River watershed hydraulic structures is founded on system planning, strengthening IoT development, deepening big-data mining, and promoting smart applications and intelligent management, transforming safety control away from the single reliance of the past on experience-based analysis and monitoring by personnel, into data-driven based smart management. Focus is centered around all-around information perception, comprehensive interconnectivity, deep data mining, and intelligent decision-making support.

- **All-around information perception**: Based on traditional monitoring methods, and combing multi-wave velocity sounding, three-dimensional laser scanning, satellite remote sensing, and other advanced technologies, build a high-coverage, all-around, all-object, all-indicator automated real-time monitoring system, enhancing the perception capabilities of the equipment, in order to provide comprehensive basic information for the intelligent operation of hydraulic structures.
- **Comprehensive interconnectivity**: Through the support of traditional communication technologies such as optical fiber and microwave, along with the application of modern technologies such as IoT, mobile internet, satellite communications, and Wi-Fi, fully build up LAN, WAN, satellite networking, and mobile networking, the four large data transmission channels, thus assembling a "big-transmission" network to support the transmission of large amounts of data, images, videos, and other information.
- **Deep data mining**: Based on the big-data platform, merge data from various specialties and share them to form big-data related to the safety of hydraulic structures, establish data standards, carry out data governance, and create a powerful algorithm model library for deep data analysis and mining, to achieve prediction and early warning of hydraulic structure safety.

© The Author(s) 2025

Y. Tu, *Management of Hydropower Enterprises*, Water Resources Development and Management, https://doi.org/10.1007/978-981-97-5584-4_7

- **Intelligent decision-making support**: Based on multi-element historical scenario similarity pattern matching, quickly generate the linked solution of *immediate state monitoring → scene simulation → handling measures response*. Establish an expert knowledge base for hydraulic structures, establish a logical reasoning algorithm, and compose a comprehensive recommended assessment solution for multi-objective control of hydraulic structures, in order to provide decision-making support for intelligent safety control for Dadu River hydraulic structures.

Taking dam operation management as an example, the Dadu River company has constructed a safety risk control and management operation system for dams in the watershed by researching the hazard-causing and failure-causing risk factors under the special working conditions for dams in a cascade group, analyzing the mechanisms of hydraulic structure failures and catastrophes, as well as the coupling mechanisms between risk factors. This is a safety risk smart control and management platform for the cascade dam group, with classical features such as data reliability analysis, multi-source information fusion, automated risk anticipation, and early warning response regulation. The platform achieves comprehensive smart acquisition, identification, exchange, integration, and analysis of multi-source data such as dam monitoring, water conditions, facility operating status, environment, boundary information, etc. It has core technologies such as smart identification of monitoring data anomalies, real-time risk anticipation, and comprehensive annual assessment, along with multi-regulation of engineering measures and management cooperation. It has functions such as real-time evaluation of dam safety risk, comprehensive assessment of operational status, safety risk warning, and responsive decision-making, and realizes dam safety risk smart control with the classical features of automated anticipation, autonomous decision-making, and self-evolution. The overall approach to dam safety intelligent control is shown in Fig. 7.1.

7.2   Key Technology

7.2.1   High Precision External Deformation All-In-One Smart Monitoring Technology Set

With the rapid development of today's computing technology and the continuous updates to observational technology, automated monitoring has replaced manual monitoring, and this has become the development trend for engineering safety monitoring. Modern computer technology, network technology, software engineering technology, engineering safety monitoring, and feedback monitoring technology have been successfully utilized across the various domains of engineering safety monitoring. Among the numerous items subject to safety monitoring, external deformation is an important indicator to assess the safety of hydraulic structures, able to straightforwardly reflect the safety status of hydraulic structures. Because of this, the

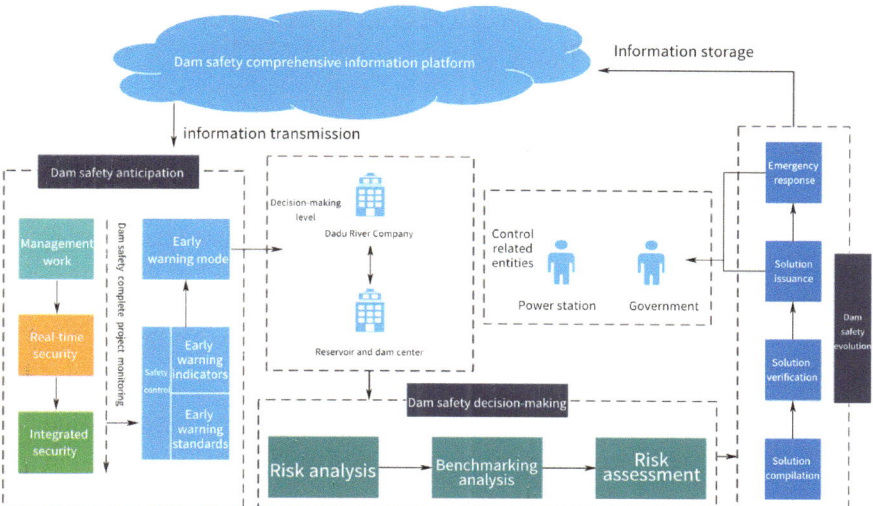

Fig. 7.1 Overall approach to dam safety intelligent control

question of how to go about automation for external deformation, and the need, is becoming more and more conspicuous and urgent. Traditional manual monitoring, however, has shortcomings in several aspects: First, there are a relatively large number of monitoring points for external deformation in the area around a dam, resulting in a high workload for manual observation, a long observation cycle, and high safety risks. There is an inability to grasp the working status of the object being managed in a timely fashion, and no way to satisfy the requirements for safe operation. Second, observation data collected via manual observation must be input into computer by hand, and manually organized and calculated. The work is rather intense, and if, in the process of organizing or doing calculations, problems are discovered in the original raw observational data or in the calculated results, then the only option is to return to the site to carry out the observations again, with very low work efficiency. Third, during operation when water level at the hydro-junction is rising or falling or at a high level, as well as during floods, earthquakes, and other extreme work conditions, the hydro-junction requires even closer observation, the observation work on site will be more tense and more labor-intensive, with a need for a greater number of personnel. Since data from manual observation makes its way into the database relatively slowly, the observation data does not get processed in a timely manner. The different levels in the enterprise hierarchy are not able to ascertain the working status of hydraulic structures in a timely manner, seriously affecting watershed flood control command decision-making and flood regulation.

In view of this, after years of research, the Dadu River Company has developed an automated system for observational monitoring that is highly accurate, full-functioned, stable, reliable, and easy to use. The system is primarily composed of the following technologies.

7.2.1.1 Baseline Self-Calibration and Meteorological Fusion Correction Technology

Any deformation monitoring system must consider the influence on distance and angle measurements of factors such as meteorological conditions, errors due to the curvature of the earth, errors from atmospheric refraction, and the long-term horizontal stability of the instruments during monitoring over time. For these factors, the traditional methodology is to carry out correction via empirical formulas for meteorology, but this approach is affected by nonlinear causes such as meteorological representation error, so that the precision of the results after adjustment is limited. In view of this, the Dadu River Company crafted a baseline self-calibration and meteorological fusion correction method.

The main approach behind this method is that when laying out the monitoring reference network for the site being monitored by the automated deformation monitoring system, select and establish some stable calibration reference points that can cover the monitoring area, and which have similar characteristics as the working reference point. The angle relationships between these stable calibration reference points, covering the entire monitoring area, and the deformation monitoring working reference point provide known conditions for automatic deformation monitoring correction or elimination of many of the errors mentioned above. During actual measurement, when at the working reference point making observation of the deformation point, the known calibration reference points used for correction are also observed according to direction, order, and elevation grouping. At this time, there will be inconsistencies between the observed values of the calibration reference points and their original known values. These inconsistencies are caused by the collective influence of the many factors, mentioned above, at a given time.

Because the amount of time needed for automated observation with the measuring robot is very short, observation values obtained at the deformation monitoring point and at the calibration reference point can be seen as being from the same relative time span. The collective impact of these calibration reference points will also in the same way, but with different magnitude, produce similar effects on other deformation monitoring points. Thus, the correction value obtained from known edges and known points can be assigned to the unknown deformation monitoring point according to mathematical model, with side length, direction value, or time difference as an argument, in order to greatly eliminate the measurement errors caused by the many factors mentioned above.

Given that the automated robotic measurement deformation monitoring system employs unidirectional observation, before using baseline self-calibration correction, described above, the observation results are first meteorologically corrected, i.e., by measuring the temperature T, air pressure P, and humidity H at the operation site, then carrying out baseline self-calibration of meteorological correction residuals according to specific meteorological correction formulas.

Using the actual observation data (horizontal angle, vertical angle, and slope distance) from the measuring robot measurement station to a calibration reference point (for which the covered monitoring area and the three-dimensional coordinates are known), the difference between the actual observation data and the coordinate inverse data for the measurement station to the calibration reference point is calculated, obtaining the meteorological correction value (related to temperature, and barometric pressure) and refractive index (related to air vertical density) for the direction of observation from the measurement station to the calibration reference point, then the temperature gradient field and air vertical density field model for from the measurement station to the monitoring area can be reconstructed. Subsequently, based on the outline coordinate of the monitoring point, the meteorological correction value and refractive index for the measurement station to the monitoring point can be calculated by using an interpolation algorithm. Finally, the correction of the original observation data (horizontal angle, vertical angle, and slope distance) from the measurement station to the monitoring point is completed.

The baseline self-calibration algorithm can complete the correction of the original monitoring point observation data (horizontal angle, vertical angle, slope distance), especially the inverse-calculation and correction of refractive coefficients, and solve the problem of the impact of large meteorological changes when using one-side triangulated elevation to carry out vertical deformation monitoring. The algorithm can also improve the precision of one-side triangulated elevation monitoring while correcting the model.

7.2.1.2   All-In-One Smart Measurement Station Platform for Deformation Monitoring

For automated external deformation monitoring systems, the measurement station is the key to the precision and reliability of monitoring data. Automated external deformation monitoring systems generally adopt methods such as polar coordinate difference or forward intersection to monitor the deformation of hydraulic structures. But no matter what method is used, the measurement station is the foundation to ensure the accuracy and reliability of the measurement data. For one thing, the three-dimensional coordinates of the measurement station are the basis for resolving the amount of displacement of the monitored object, so the long-term steadfastness of the station itself must be ensured. At the same time, as the measurement station is generally located in the field, and has precision instruments such as measuring robots installed, the various equipment needs to operate out in the field for a long period. Therefore, how to ensure the safety of the precision instrumentation, and how to properly protect against problems like theft, vandalism, and lightning is a key to ensure the accuracy and reliability of measurement data, and is also a key issue for consideration during the design of an automated external deformation measurement station.

Based on the importance of safety control in engineering projects, and in order to make targeted improvements to traditional measurement stations, the Dadu River Company has developed an integrated all-in-one smart measurement station for deformation monitoring, with large field-of-view and integrated control of multiple types of high-precision instrumentation. Combining in information technology,

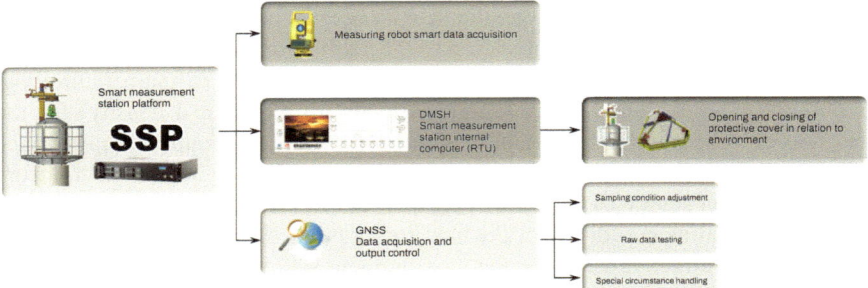

Fig. 7.2 All-in-one integrated smart measurement station platform architecture

smartness, and IoT, etc., the station achieves concentric mounting and integrated management of the precision instrumentation, and can automatically control the smart perception of external environmental conditions and control the storage environment for its precision instrumentation. The station expands the scope of observable field-of-view for the measurement point area of a single measurement station, and can smartly select the best timing to carry our measurement and then automatically open or close the observation portal of the measurement station (via remote timing or in real-time), working with the measuring robot, which has automated identification functionality, to complete the arranged deformation monitoring tasks, and replacing manual human observation operations. This improves measurement precision and increases promptness of information return. The all-in-one integrated measurement station solves the difficulties in protecting precision instruments located in the field for long periods, reduces investment of enterprise resources along with safety risk for operators in the field, and provides smart solutions and innovative ideas for deformation monitoring in engineering projects. Platform architecture is shown in Fig. 7.2.

The integrated all-in-one smart measurement station platform realizes five functions: Concentric installation and integrated centralized management of instrumentation, large field-of-view station design, remote automated opening and closing control of the observation portal along with field protection of instruments, monitoring of station status and smart regulation of temperature and humidity, and smart perception and judgment of the external observation environment.

The key to achieving these five functions is the measuring robot measurement station all-in-one integrated control system. This system creates a linkage between the observation task of the measuring robot, and the opening and closing of the portal of the measurement station, and embeds this linkage into the on-line deformation monitoring software system, accomplishing remotely timed and real-time opening and closing of the observation portal, monitoring of measurement station status, and smart regulation of measurement station temperature and humidity. Each control component of the measurement station system is integrated and connected to the measurement station's internal computer (RTU), and manual or automatic control happens via this local computer or via remote computer.

7.2.1.3   High Precision External Deformation Smart Monitoring System

The Dadu River Company has developed a dam deformation smart monitoring system based on the measurement principles of the baseline self-calibration and meteorological fusion correction method. This system automatically carries out all-in-one implementation of the complete monitoring task, from the formulation of dam deformation monitoring tasks, to collection of monitoring point data, and finally to result analysis for the monitoring task, achieving real-time data reporting, prompt data analysis, automated generation of reports, automated control of equipment, and automated alerts for anomalies, etc. Simultaneously, dams with different conditions or multiple dams together can utilize the internet to form group type management.

The system mainly includes: Smart dam deformation monitoring stations, smart station control system (introduced earlier), dam deformation monitoring data acquisition system, dam deformation monitoring central data platform, and dam deformation monitoring analysis system. At the same time, in order to meet the requirements for precision and timeliness in the automated monitoring of three-dimensional surface deformation and to circumvent the deficiencies of traditional monitoring methods, and in overall consideration of technical and economic targets, the Dadu River Company further introduced a high-precision automated monitoring solution for three-dimensional surface displacement, primarily employing monitoring via measuring robot, and supplemented by measurement via GNSS satellite positioning. The automated dam deformation smart monitoring system, using high precision measuring robots based on an internet platform bringing together computer technology, automated measurement technology, spatial technology, sensor technology, and communication technology all into one, integrates data functions like acquisition, real-time output of results, and alerts for anomalous values for hydraulic structure external deformation monitoring data, forming a relatively complete monitoring system. Additionally, the effectual combination of information management analysis software and prepared extra interface capacity, accomplishes rapid early warning and emergency handling decision-making for automated monitoring.

The data retrieval subsystem of the dam deformation smart monitoring system is a comprehensive management system that automatically completes collection, storage, and transmission of observation point data following established observation plans or at independently selected best measurement opportunities. Its orderly operation is the prerequisite for providing fundamental and valid data for dam deformation monitoring safety analysis. The system contains seven main functional modules, including system set up, observation point management, monitoring plan, data import, data export, access monitoring, and initialization.

The central data platform is the core of the entire automated dam deformation monitoring system, bringing together functions like system set up, monitoring task formulation and issuance, monitoring data receiving, look up, storage, import, and backup, all into a single important platform. The central data platform is the basis for operation of the acquisition platform and post-processing platform. The platform is first-hand material for monitoring data analysis, and a platform for the storage, look up, and management of all original raw data. The central data platform is connected via network to each retrieval platform to facilitate the receipt of acquired monitoring data from each retrieval platform. The network topology is shown in Fig. 7.3. The basic functions of the central data platform include: Data reception and look up, dictionary definition management, retrieval mode and plan, monitoring task issuance, importing retrieved data, system authority management, system initialization, and off-site data backup.

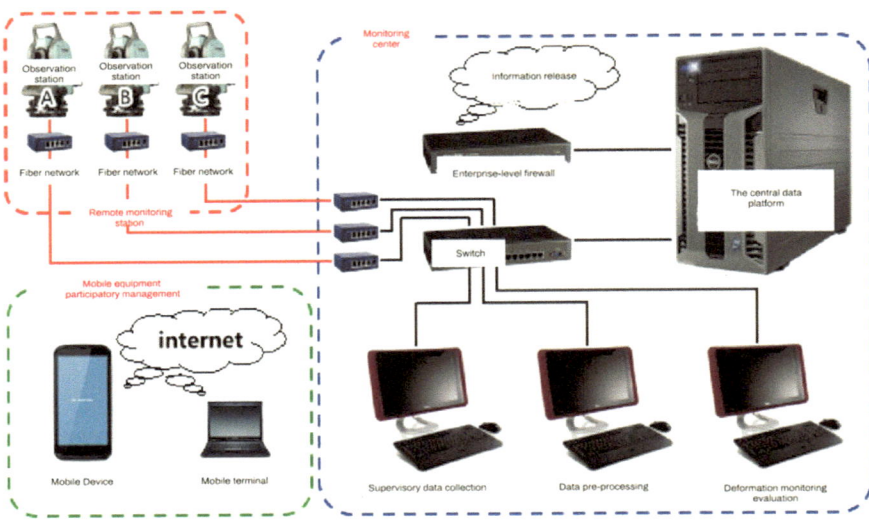

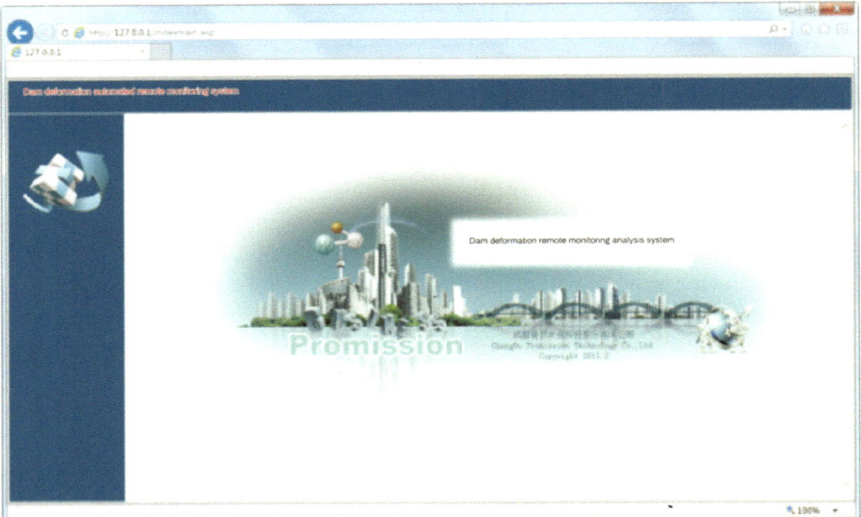

Fig. 7.3 Dam deformation monitoring central data platform topography, and system interface

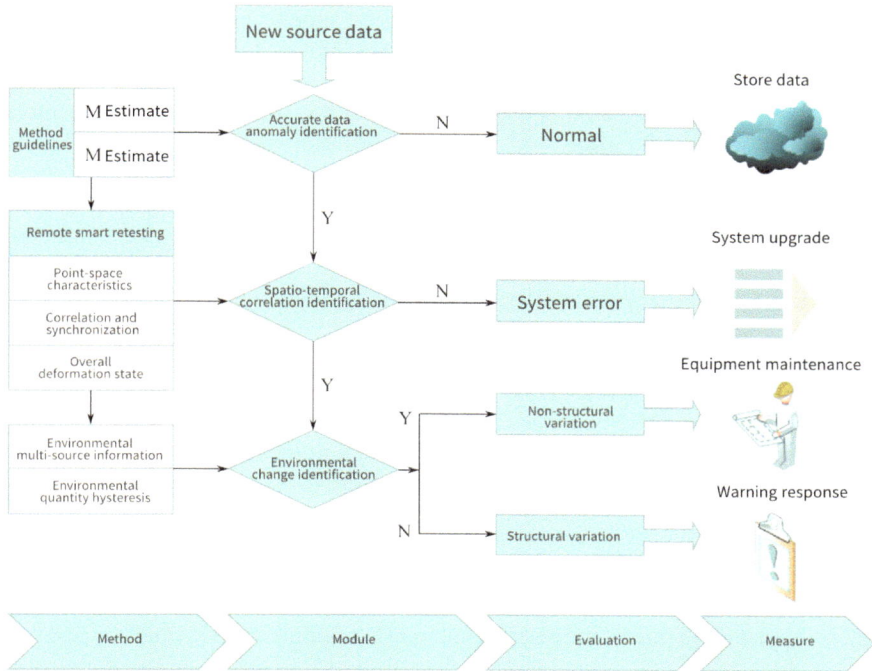

Fig. 7.4 Complete set of technologies for on-line identification of anomalies in safety monitoring data for the Dadu River watershed

The baseline self-calibration and meteorological fusion correction method greatly reduces the influence of external uncertainties such as meteorological conditions on measurement precision. The TPS\GNSS fusion measurement solution achieves continuous retrieval of real-time monitoring information under severe weather and special working conditions. Independent design, and development of an all-in-one measurement station with superior station structure, achieve a large increase in the scope of observable field-of-view for the measurement point area. With an integrated product design concept, the integrated measurement station brings together meteorological sensors, cabinet air conditioners, electromechanical cylinders, measuring robots, retroreflectors, GNSS antennas, video surveillance systems, and various photoelectric conversion equipment, achieving the integrated management of multiple types of instruments and equipment, and providing innovative conceptualizations for the construction of automated monitoring system equipment. The measurement point reflector and GNSS antenna are arranged around the same axis, and the instrument height is set to a fixed length, reducing the measurement error. The measurement point reflector and the GNSS antenna both use the same size protective cover and narrow reflector observation portal design, and are bolted internally, with specialty tools needed to open or close, effectively solving the problem of protection and anti-theft for field measurement instruments in the dam area, and laying a solid foundation to ensure the quality of monitoring data. The development and application of an all-in-one integrated control system for measuring robot monitoring station, which goes hand in hand with survey robot automated observation, solves the problem of all-weather real-time startup of the robot measurement station, and achieves the timed or manual opening and closing of the observation portal, with the observation portal opening and closing according to the measurement task to protect the measuring robot from damage. Equipment cabinet air conditioning, meteorological instruments, and video surveillance systems are installed inside the measurement station, along with a control system for the electromechanical cylinder, and the development of the measuring robot measurement station all-in-one integrated control system achieves control of the electromechanical cylinder, video surveillance, monitoring of internal temperature and humidity, and smart automated startup and stop of cabinet air conditioner, as well as smart adjustment of temperature and humidity. The integrated measurement station that can smartly select the measurement time interval, and can smartly abandon the measurement task upon encountering strong wind and heavy rainfall, in order to protect the precision instruments in the station and improve the overall quality of the measurement data is introduced. In relation to the timing for opening the protective cover for observation, in addition to considering rainfall, wind speed, and other conditions, the measurement station also considers changes in the temperature gradient, so that the measuring robot automatically selects the period of no wind, no rainfall, and stable temperature and pressure to carry out observation, thereby minimizing the influence of meteorological conditions on observation precision, and ensuring the accuracy and reliability of the monitoring data.

To sum up, the Dadu River Company has achieved integrated smart monitoring, bringing together such various functions at environmental identification, automatic opening and closing, etc., thus reducing or even circumventing impacts of rainfall, wind speed, visibility, and temperature gradient on measurement precision, with capability to smartly identify the best observation time in order to obtain high confidence measurement data, while also ensuring real-time and continuous data.

7.2.2 Combined Multi-method Above- and Under-Water 3D Digital Defect Detection Technology

Hydraulic structures are an important component of dam safety control, and their structural–functional design is a key to the long-term safe and stable operation of the dam. After the dam is built and water has been impounded, the underwater structures are submerged all year round, and under the effects of the complex stream environment. There will be different degrees of siltation, material deterioration, functional degradation, and other phenomena, and the defects are characteristically difficult to discover, difficult to deal with, occur suddenly, and lead to serious consequences. Conventional means of dam safety monitoring is to try to get an understanding by sampling certain points, then progressively building up bit-by-bit to reflect the operation of complete hydraulic structures. With this it is only possible to achieve status monitoring for crucial locations, and full coverage dynamic and quantitative monitoring and analysis is lacking. In order to solve these issues, the Dadu River Company carried out research and application, developing the following four technologies:

(1) Multibeam deep-water marine underwater detection technology.
(2) Underwater submersible vehicle carried camera, two-dimensional and three-dimensional sonar close-range camera and scanning technology.
(3) 3D laser (station-mounted, ship-mounted, and airborne) full-coverage scanning and high mobility technology.
(4) Shallow stratigraphic profiler geological defect detection technology.

Through the complementary strengths and joint application of multiple technical methodologies, the technical problems of 3D digitalization of hydropower station hydraulic structures have been solved.

In view of the challenges that exist for traditional inspection of hydraulic structures (such as dam-forebay, flood discharge, and tailrace structures, etc.), multibeam detection technology and underwater unmanned submersible vehicle inspection technology are employed to carry out nondestructive inspection of hydraulic structures, overcoming the challenges of traditional manual underwater inspection. At the same time, multibeam detection technology achieves accurate positioning, quantification, and 3D point cloud data visualization of underwater hydraulic structure defects, while underwater unmanned submersible vehicle collects images of defects, the results of the two approaches corroborating each other.

For hydropower station hydro-junction areas, traditional manual inspection is greatly affected by problems that exist in quality of technical personnel, low inspection precision, and low efficiency. To address this, the Dadu River Company utilizes 3D laser (station-mounted, ship-mounted, and airborne) scanning technology to carry out regular digital inspection on the dam (including galleries), flood discharge tunnels, and other constituents, achieving digital inspection and quantitative comparative analysis of these areas.

Full-coverage scanning survey of the dam forebay river channel and both sides of the hydropower station dam is carried out through joint application of multibeam detection technology and 3D laser scanning technology, combined with RTK GPS high-precision positioning technology, gaining accurate above- and under-water three-dimensional digital results for hydropower station hydro-junctions. The state and distribution of sediment in the dam forebay area is comprehensively captured, and the operation status of hydraulic buildings comprehended, supporting later quantitative analysis with accurate data.

In addition to the technical research applications above, the Dadu River Company also brings together hydropower station design drawings, past inspection results, and other relevant materials, along with comprehensive dam structure 3D data visualization technology, and carries out integrated analysis and research application of the 3D digital results, so as to provide reliable data support for work arrangements related to hydropower station hydraulic structure safety management and other domains.

At the same time, the integration of multibeam detection technology, underwater unmanned submersible inspection technology, and three-dimensional laser scanning technology, comprise a new technological means of above- and under-water integrated monitoring and inspection, establishing a high-precision, full-coverage underwater 3D digital results library for hydraulic structures, realizing the accurate positioning and quantitative analysis of underwater defects for hydraulic structures, solving the technical problems for inspection of hydraulic structures in deep, turbid, and moving water environments, and breaking through the technological bottlenecks of traditional inspection by manual human diving or robot diving. The integration of 3D design and modeling technology achieves digital inspection and dynamic monitoring of large- and medium-sized hydropower project dams, galleries, and flood discharge tunnels, and quantitative analysis of the operation status of these structures saves much human resource cost. With the integration of high-precision scanning point cloud coordinate data and dam 3D model data, a dam's abnormal zone (including location, size, and area) and above-and under-water terrain information can be dynamically displayed and extracted in real-time. This can accurately guide the repair of hydraulic structure underwater defects, and opens an interface with the watershed dam safety risk smart control and management platform, which lays the technical support for deep mining and application of dam safety monitoring data.

7.2.3 Technology for Cascade Dam Group Real-Time Safety Risk Anticipation

The formation of the real-time smart safety risk anticipation system for the Dadu River dam group originates first from the high reliability of the data that are part of the evaluation, and then, with the reliability of the data assured, real-time smart assessment of dam operation safety can be made.

7.2.3.1 On-Line Identification of Safety Monitoring Data Anomaly

In the early days, evaluation of monitoring data reliability by monitoring data analysts mainly relied on graphing the curve of effect volume versus time to see whether there were obvious outlier points. Then, if there were outliers, analyzing whether the measured values were anomalous based on a combination of engineering experience and inspection circumstances. The graphing method is based on the engineering experience of the analysts looking at the monitoring materials. The method has no clear determination criteria, a great amount of subjectivity, and is very difficult to automate. With the change of dam safety management from extensive management to technology-based, risk-based management, watershed power stations gradually started building a dam safety on-line monitoring system based on modern information technology, computer technology, and dam engineering theory. The amount of monitoring data has increased dramatically, and real-time rational evaluation of the reliability of data from new sources is a prerequisite and important assurance for the normal operation of the on-line monitoring system.

In response to the shortcomings of 3σ criterion, the mathematical modelling method, and the uncertain-numbers method, the Dadu River Company has for the first time introduced a set of on-line identification technologies for safety monitoring data anomalies, bringing together accurate data anomaly identification, spatio-temporal correlation identification, and induced environmental change identification. New $3S_T + D$ judgement criteria are employed for on-line real-time identification of anomalous sudden changes in measurement values, afterwards triggering start of environmental correlation analysis of water level, rainfall, regional earthquakes, near-area disturbance, etc., filtering and eliminating sudden variations induced by changes in environmental quantity. Advanced anomaly measurement points trigger initiation of high-precision multi-dimensional spatial model simulation, analyzing the spatiotemporal distribution characteristics and patterns for same-type measurement points along different dimensions such as line, plane, and space. Additionally, at a suitable time, remote retesting is smartly initiated, and abrupt changes induced by anomalous system measurement values are eliminated through feedback calibration, so that irregular measurement values arising from actual changes in structural safety state are finally accurately identified and early warning is triggered automatically. Through this technology, the rate of missed and misjudged data anomalies can be reduced by more than 96% compared with traditional identification approaches.

The complete set of technologies for on-line identification of anomalies in safety monitoring data is shown in Fig. 7.4.

7.2.3.2 Safety Risk Real-Time Smart Anticipation

Traditional judgment of the site-measured state of hydraulic structure safety mainly happens with the help of engineering experience and the establishment of mathematical models for individual measurement points, and for individual items. This method has certain limitations in reflecting the overall safety of the dam in terms of structural state. In order to make a complete and rational assessment of the operational safety and structural state of a dam, it is necessary to comprehensively consider the monitoring points and effect size of each part of the dam, then use mature and appropriate analysis methods to make an overall evaluation. The Dadu River Company gradually launched on-line real-time monitoring of dam safety risks, achieving smart management with real-time mastery of dam safety risks, improving the level of scientific decision-making and the capability for safety assurance.

Based on the guidelines, the Dadu River Company has constructed an on-line smart rapid comprehensive assessment model for dam operation safety risk, which realizes on-line rapid appraisal of dam operation safety risk ranking. For projects that have achieved real-time assessment of key indicators of dam operation safety, the assessment model can quickly and comprehensively carry out appraisal and produce a comprehensive appraisal ranking for dam operation safety risk based on actual measurement data or real-time assessment results. At the same time, a fuzzy entropy-weighted annual comprehensive assessment model for dam operation safety was constructed, and a fuzzy clustering comprehensive assessment method and an extenics-based comprehensive assessment method were introduced both for dams with abundant periodic inspection data and for those lacking periodic inspection data, respectively. This can effectually avoid the problems with overly deterministic indicators and conflicting incompatible indicators, and can more objectively reflect the annual operation safety characteristics of dams, providing clearer and more credible decision-making support for dam safety management. The structure of the comprehensive assessment system is shown in Fig. 7.5.

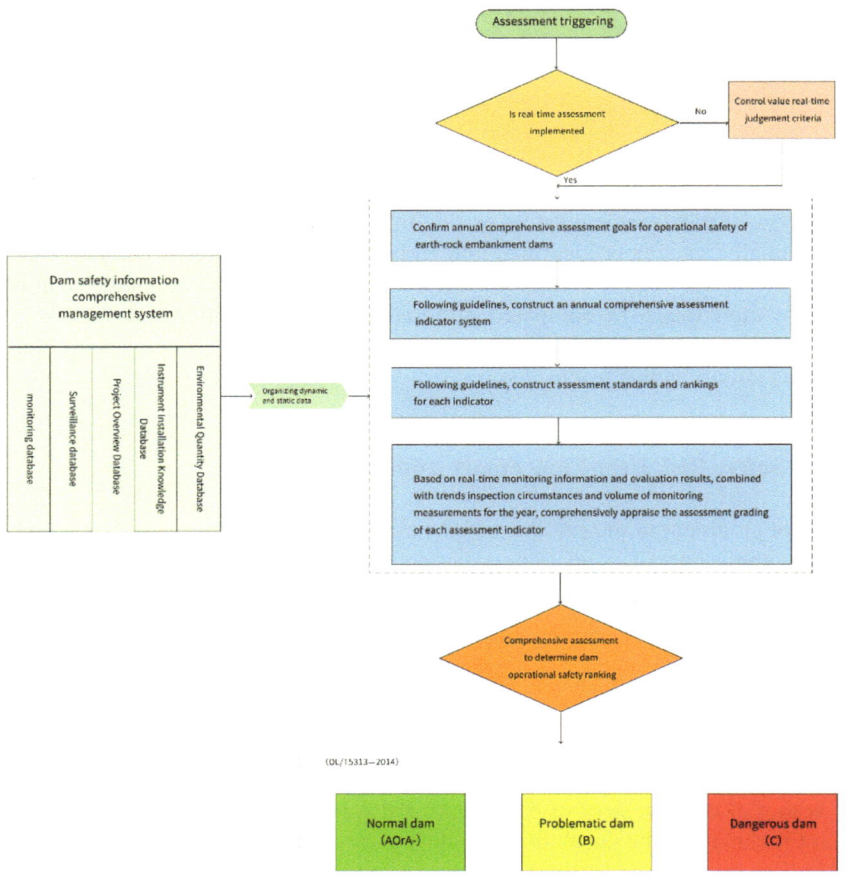

Fig. 7.5 Structure of dam operation on-line comprehensive assessment system for safety risk

7.3 Case Application

7.3.1 Successful Response to the "8–20" Mudslide

7.3.1.1 Basic Conditions

On August 20, 2019, an exceptionally large flash mudslide occurred in Wenchuan County, Aba Prefecture, Sichuan Province, China, causing a number of hydropower stations in Wenchuan County to be impacted to varying degrees, including one hydropower station for which the flood discharge channel was blocked by mudslide sediment, the dam's working power supply was destroyed, and the dam was overtopped. With emergence of these dangerous circumstances, 119 people were trapped for a time upstream of the dam, while the dam faced the risk of destabilization and collapse, seriously threatening the lives of about 5,000 people downstream.

7.3.1.2 Emergency Response Situation

During the emergency response period, the Dadu River Company rushed dam safety monitoring specialists and technical personnel to the scene to support the rescue and relief work. These personnel were divided into three working groups for aerial survey by UAV, measurements by unmanned surface vehicle (USV), and barrage dam safety monitoring.

UAV Aerial Survey Clarifies Disaster Impacts in Blind Zones

On August 21, 2019, the drone aerial survey team used UAV aerial survey technology to go deep into the scene of the affected hydropower station. The team developed a detailed UVA survey program for the limiting conditions at the scene, where road transport was interrupted and personnel could not get to the dam site, and were only able to see dam conditions from a distance of 100 m away. Straight away, the team completed its first all-around and multi-angle site image data collection post-disaster, and afterwards completed dozens of successive UAV aerial surveys, gathering nearly 20G of high-definition image data of the disaster site, providing robust support for emergency response command in formulating an early emergency rescue solution, solving the challenge of not being able to see actual conditions at the hazard site. UAV aerial imagery of the complete hazard scene, and conditions of key component areas, is shown in Fig. 7.6.

Fig. 7.6 UAV aerial imagery of the complete hazard scene, and conditions of key component areas

USV Measurements Reduce Safety Risks Downstream

On August 30, 2019, from 14:00 to 16:00, the USV measurement work team chose a zone with good GPS signal on top of the dam's diesel generator room to erect a reference station and used an USV to carry single-beam depth sounding to carry out underwater topographic measurement in the more stable water flow area in power station dam forebay area (as shown in Fig. 7.7). The team obtained 3D coordinate information for about 3000 measurement points underwater, thus obtaining results for reservoir volume capacity and cumulative sediment siltation volume. Through the application of USV-carried depth sounding technology, sediment siltation elevation data for the dam forebay is accurately obtained, and additionally, reservoir capacity data for different water level elevations can be calculated, allowing understanding of the patterns of sediment siltation in the dam forebay area, and providing technical substantiation for judgements regarding whether reservoir water discharge will create secondary disasters downstream.

Barrage Dam Safety Monitoring Confirms Dam Stability

In order to accurately monitor the safety and stability of the barrage dam and ensure the personal safety of the rescue personnel on site, from August 30 to September 3, 2019, the barrage dam safety monitoring team researched and formulated an emergency monitoring solution for the actual site conditions. The team laid out 13 deformation monitoring points and three mounted mini-sensor monitoring points on the top of the dam (as shown in Fig. 7.8), and conducted dam top deformation monitoring 59 times using high-precision smart measuring robots. The mounted mini-sensors were used to continuously monitor the vibration and tilt of the dam top, and to configure the early warning threshold for monitoring indicators. Accurate analysis of the safety monitoring results showed that dam monitoring indicators had changed very little, and the dam was considered safe and stable, with no dam failure risk, providing

Fig. 7.7 Dam forebay area underwater measuring via USV-carried depth sounder

strong support for people downstream to resume production and their personal lives at an early time.

7.3.1.3 Primary Results Attained

The use of advanced UAVs, USVs, measuring robots, and other perceptive equipment to clarify the current situation of the hydropower station under conditions of flood-obstruction transportation routes, the continuous tracking of the safe and stable operating status of the dam during the flood inundation period, analysis of the of dam flood failure evolution and potential downstream degree of harm, the provision of reliable monitoring data to the disaster rescue and relief command, all provided forceful support for controlling the safe operation of the dam, and for the protection of the rescue team members and their equipment. Additionally, these also provided technical support for evacuated people downstream of the dam to return to their homes, resume production, and resume their lives, effectually eased the pressures on

Fig. 7.8 Layout of the barrage dam safety monitoring facilities

emergency rescue personnel while handling dangerous situations, and contributed to the successful handling of a dangerous situation.

7.3.2 Accurate Identification of Underwater Structural Defects at Tongjiezi Power Station

7.3.2.1 Basic Conditions

Tongjiezi hydropower station is located within Shawan District, Leshan City, Sichuan Province, downstream of the Dadu River, and has been operating for many years since it was commissioned in October, 1993. In order to ensure the good operation of water retaining and energy dissipating hydraulic structures, inspection of the dam and downstream energy dissipation and scour prevention facilities was carried out on at planned regular periods since the commissioning of the dam. Previously, single-beam spot detection had always been used, combined with manual diving, which was limited by small detection range and low precision.

To solve this problem, the innovative joint use of multibeam detection technology and shallow stratigraphic profiling technology achieves comprehensive analysis and comprehension of near dam underwater siltation and hydraulic structure defect conditions, and provides technical guidance for the subsequent operation of hydraulic structures.

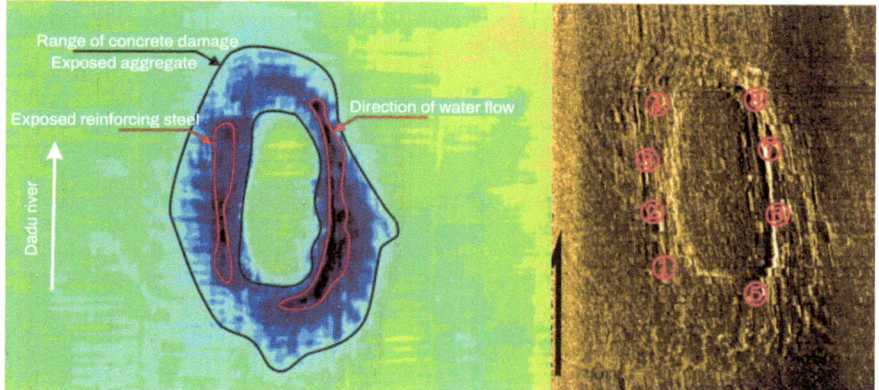

Fig. 7.9 Tongjiezi hydropower station dissipation pool underwater inspection results and underwater photography results

7.3.2.2 Application Results

In 2017, multibeam detection technology and shallow stratigraphic profiling technology were applied to the dissipation pond below Tongjiezi dam. It was discovered that there was a zone with a certain scale of partial concrete erosion in concrete bottom of the forebay to the dissipation pool, shaped approximately elliptical in the direction of water flow, the length of the long axis being about 70 m, and the length of the short axis being about 35 m. Perimeter erosion was found to be more serious, accompanied by the occurrence of exposed aggregate and exposed steel reinforcement, which had impact on the structural integrity of some of the pool forebay bottom, and which would endanger flood discharge safety if it developed further.

In view of this, in 2018, the Dadu River Company conducted further examination of the underwater inspection results, and based on the concise and clear underwater inspection results, a rational underwater repair solution was formulated to repair the defects of the dissipation pool and ensure the safe and stable operation of the flood discharge and energy dissipation structures.

Given the importance of this key structure for flood discharge and energy dissipation at Tongjiezi hydropower station, the discovery and timely remedy of this relatively large-scale defect prevented the further expansion of defect in the concrete bottom of the pool, which guaranteed the flood discharge capacity and regulation capacity of Tongjiezi Hydropower Station and safeguarded the lives and property of downstream residents.

Tongjiezi hydropower station dissipation pool underwater inspection results and underwater photography results are shown in Fig. 7.9.

Chapter 8
Intelligent Protection of Watershed Ecosystems

8.1 Approach and Goals

With the gradual development of hydropower on the main stem, the Dadu River Company has paid increasing attention to the impact of hydropower development on the ecological environment of the watershed, and to the restorative effects of related environmental protection measures on the ecological environment. However, limited in the past by both environmental monitoring methods and by the technological level of the monitoring equipment employed, the monitoring and perception network did not make up a complete system, and interconnectivity for ecological environment data had not been achieved. Although some on-line monitoring systems did help in environmental management, yet almost all ecological and environmental risk identification and emergency response decision-making still relied on worker experience. There was a lack of environmental protection intelligent data analysis and risk early warning for hydropower development. Environmental protection risk identification and control had not reached the level of intelligence across all aspects and all processes.

Given these issues, and in consideration of the ecological protection needs of the Dadu River watershed and of the current protection status of sensitive subjects, overall planning and construction of an ecological environment perception network was carried out, eliminating data barriers. An ecological environment protection platform with intelligent perception, dynamic assessment, real-time warning, and assisted decision-making was created. The standardization of the ecological environment monitoring system, automation of equipment, data sharing, intelligent modelling, and intelligent decision-making support were all promoted in an orderly fashion, in order to realize dynamic assessment and decision-making support of environmental risks throughout the whole process of hydropower development.

Y. Tu, *Management of Hydropower Enterprises*, Water Resources Development and Management, https://doi.org/10.1007/978-981-97-5584-4_8

The focus of intelligent ecological environment protection in the watershed is to build an ecological environment perception network, to realize multi-source ecological environment data integration, to establish an ecological environment risk early warning system, and to build an ecological environment intelligent control model.

Build an ecological environment perception network: Employ 5G and other recent generation mobile internet technology and advanced instrumentation to acquire various types of ecological and environmental monitoring data in real-time, fully perceive the ecological and environmental conditions of the watershed, and build an ecological environment perception network. For those monitoring variables for which it is currently difficult to achieve smart perception, extra interface capacity for expansion has been set aside, and upgrades can be made as the technology matures.

Achieve the integration of multi-source ecological environment data: According to the structural type, spatial distribution, sampling frequency, and other characteristics of data related to hydropower ecological environment protection, integrate the environmental monitoring data of each cascade hydropower station, as well as existing related data held by ecological environmental protection management departments. Establish a unified multi-source ecological environment database, and transmit and store real-time watershed ecological environment monitoring data in the big-data center.

Establish an ecological environment risk early warning system: Based on current needs for ecological environmental protection work in hydropower development, strengthen mining and application of big-data, establish a risk early warning system, and for various types of ecological environmental risk, accomplish current condition assessment, trend analysis, and graded early warning.

Build an ecological environment intelligent control model: Transform ecological environmental protection management experience into knowledge, and put together a knowledge base for ecological environmental protection management. Combine analysis and mining of real-time monitoring data into ecological and environmental risk control and decision management for hydropower development. Through the analysis and judgment of the knowledge base, provide accurate and effective support for the intelligent control and decision consultation for ecological environmental protection on the Dadu River.

8.2 Key Technology

8.2.1 Smart Monitoring of Flow Field and Image Recognition of Target Fish Species at Fish Passage Facilities

After a reservoir and hydroelectric power plant are built, under typical conditions, fish and other aquatic organisms cannot travel freely up or down in the river due to the obstruction created by the dam. This has an impact on fish breeding, feeding, overwintering, and other important living and survival requirements. Fish passage facilities are man-made channels that allow fish to pass through locks, dams, and other barriers, and are a key measure to restore river connectivity, and can include items such as fish passages, fish locks, fish lifters, imitation natural bypass fishways, and fish collection and transportation boats. At present, individual fish passages have been completed at both Zhentouba (Level I) and the Shaping (Level II) power stations.

8.2.1.1 Fishway Flow Field Monitoring

Respective monitoring data of the inlet area of the fishway below the dam, of the outlet area of the fishway above the dam, and of the flow field of each pool of the fishway are all important foundations for assessing whether hydrodynamic indicators, in practice, match original design indicators, and for improving the operational effectiveness of the fishway. Monitoring of the flow fields at the inlet of the fishway below the dam and at the outlet of the fishway above the dam is carried out by an unmanned surface vehicle with an acoustic Doppler current profiler (ADCP), which navigates along pre-planned routes automatically. Flow field monitoring for each pool chamber of the fishway is carried out via lattice measurements using 3D single-point acoustic Doppler velocimetry (ADV). Additionally, video equipment is used to record the flow patterns in each pool of the fishway, including large and violent eddies, surges, hydraulic jump areas, and backwater areas, to explore quantitative analysis of the fishway flow field through image recognition. Observation of the flow field of the fishway under various working conditions, and comparative analysis under the various conditions of the differences in flow velocity and flow patterns of the fishway inlets, outlets, internal pool chambers, provides scientific basis for improving the operation of the fishway.

8.2.1.2 Monitoring of Spatial and Temporal Distribution of Fish in the River Segment Below the Dam

Spatial and temporal distribution of fish in the section of the river below the dam are investigated using an echo fish detector along with both cross-sectional fixed-point and moving sonar detection methods. Cross-sectional fixed-point sonar detection primarily analyzes the pattern of fish groups traveling up and down stream, along with the size of individual fish. Moving sonar detection primarily analyzes the spatial and temporal distribution of fish groups in the region below the dam, and obtains information on the water layers with fish distributions. At the same time, power station operating discharge rate, water level, turbine working status, etc., are all recorded, creating a cloud map of fish distribution below the dam under different operating conditions.

8.2.1.3   Image Recognition of Target Fish Species

Fishway fish passage counts employ infrared grating, with closed processing, automatically collecting data for fish numbers, fish lengths, swimming speeds, and the outline shapes of the fish. At the same time, in the fish passage observation room, an underwater image acquisition system has been installed. This system includes an underwater video camera and flash lighting, along with a water temperature sensor, and video is captured by trigger for passing fish, and water temperature is measured. Use of automated image recognition functionality based on artificial intelligence algorithms for target fish species is being actively explored.

8.2.1.4   Tag Tracking and Monitoring Inside the Fishway and Reservoir Area

Passive integrated transponder (PIT) tags are used to track and observe the behavior of target fish in pool chambers, analyze their process of swimming behavior while passing through the fishway, discover key factors affecting their passage, and come up with appropriate working and hydrodynamic prerequisites for the passage of target fish through the fish passage. RF tracking technology and video monitoring are used to record circumstances related to fish passing through the common pool chamber, resting pools, turns, and dam-crossover section. Test fish in pool chambers are observed swimming, going back and forth, and resting, with emphasis on fish passing behavior at locations such as the sites where the partition walls change form, locations where fish could not pass or transit took more time during the experiment, and at resting pools, turns, and the dam-crossover section.

8.2.2   Remote Telemetry Technology for Fish Hatchery Management and Fish Release Effectiveness

Water impoundment for cascade power stations changes the original flowing life environment, and important habitats are flooded, such as original fish spawning grounds in the river. This causes the decline of fish stocks and a decrease in genetic diversity of fish populations. Fish hatching and release is one of the principal measures to effectively mitigate the adverse effects of hydropower projects on fish stocks. Fish hatcheries have been built at Pubugou and Houziyan. As of July 2021, a total of 8,874,500 fish fry have been released from the fish hatchery at Pubugou, and 1,365,000 fish fry have been released from the hatchery at Houziyan.

8.2.2.1 Explorations in Monitoring Release Results

Results of fish release are monitored via in external marking, internal marking, chemical marking, molecular marking, ultrasonic telemetry, and other techniques. For different species, sizes, and habits of release fry, exploration is being carried out in finding suitable means and technologies for fry marking, in order to improve the success rate of marking and release, and the accuracy of result assessment.

8.2.2.2 Explorations in Assessment of Release Results

Using survey data, focused research looks at information related to stock densities for released species, population structure, feeding circumstances, as well as patterns of migration, to grasp fish behavior and patterns of change, and to scientifically assess the quantity of fishery breeding resources. Exploration of an assessment system for the results of hatching and release provide a scientific basis for decision making and adaptive management for the fish stock hatching and release program in the watershed.

8.2.3 Technology for Water Quality Dynamics Monitoring and Anomaly Warning

In order to objectively assess and understand the impact of cascade hydropower station operation on the reservoir area and on the down river aquatic environments, and to prevent and respond to sudden environmental pollution events, the Dadu River Company has built monitoring stations for water quality dynamics covering the reservoir areas and river courses of the watershed, and has established a water quality anomaly early warning platform, with functions such as water quality on-line acquisition, data storage, anomaly early warning.

8.2.3.1 Layout of Monitoring Stations for Water Quality Dynamics

Initially, eight water quality monitoring stations were constructed in the watershed, monitoring reservoirs and their down river courses, including the Houziyan power station, the Dagangshan power station, and the Pubugou power station, in order to form a relatively systematic monitoring station network for water quality dynamics in the watershed. Monitored water quality indicators are primarily turbidity, pH, dissolved oxygen (DO), total nitrogen (TN), total phosphorus (TP), chemical oxygen demand, and permanganate index. Water quality monitoring through the monitoring stations employs a combination of traditional sample measurement and on-line measurement acquisition.

8.2.3.2 Water Quality On-Line Acquisition

Taking Pubugou hydropower station as an example: A floating smart environmental monitoring platform was constructed, with emphasis on monitoring water quality conditions at reservoir inlet and outlet cross-sections. This is matched with satellite communication technology, real-time retrieval and remote transmission of water environmental monitoring data, focusing on the analysis of water quality trends in the reservoir area.

The floating smart monitoring platform uses a marine carbon steel manufacturing process, and is designed with lifting hoist rings and side pull hoist rings for easy transportation and for securing on the water. The surface of the floating platform is outfitted with the necessary monitoring equipment, and the equipment is powered with photovoltaic solar power panels paired with lithium batteries. Equipped with Beidou short message communication equipment, the platform can retrieve and report data to the server in areas of the watershed without base station network coverage. Or, paired with base station network communication equipment, the platform can synchronously report data, with a reporting interval that can be flexibly set at between one minute and 60 min.

8.2.3.3 Application of Water Quality Early Warning

The Dadu River Company has put together a set of on-line early warning applications based on water quality monitoring data. On a watershed cascade power station distribution map, for each cascade power station, it is possible to display the locations, distribution, and numerical information of water quality monitoring cross sections and pollution discharge monitoring points, as well as water quality figures for each monitoring station. At the same time, for anomalous water quality values, early warning messages and data source information can be sent in real-time, based on preset water quality threshold values. Additionally, the Dadu River Company is actively developing embedded reservoir-river coupled hydrodynamic water quality models and advanced water quality warning applications based on neural networks and other big-data algorithms, achieving functions such as watershed water quality trend forecast, pollution traceability analysis, and emergency responsive decision-making for sudden environmental events and other functions.

8.2.4 Exploration in Real-Time Monitoring of Reservoir Water Temperature and Smart Control

With the development of cascade hydropower on the Dadu River mainstem, the construction and operation of a portion of the reservoirs will have certain impacts on the spatial and temporal distribution of temperatures in the original natural river course, and will cause vertical temperature stratification of the water body, the degree of intensity showing regular seasonal patterns of change. In order to systematically comprehend the relationship between reservoir operation and reservoir water temperature structure, dam forebay water temperature distribution, and discharge water body temperature, and to fully understand the impact of hydropower plant operation on river water temperature, the Dadu River Company has carried out exploration of real-time monitoring and smart control of reservoir and downstream river water temperature. Work has already been done in on-line monitoring of dam forebay and downstream water temperature at the Pubugou, Dagangshan, and Houziyan hydropower stations.

8.2.4.1 Water Temperature On-Line Acquisition

The Dadu River Company has established a dam forebay and downstream river water temperature acquisition system using advanced Chinese water temperature measurement and reporting technology and automated instrumentation. The system has relatively high monitoring accuracy and timeliness, and is composed of on-line telemetric water temperature instruments, and a central data receiving station, as shown in Fig. 8.1.

The water temperature telemetric instruments are composed of devices such as a telemetry terminal (RTU controller), communication equipment, power supply, and temperature sensors. Based on observed on-site conditions, on-line monitoring cross-sections are laid out in typical areas of the dam forebay and below the dam. Dam forebay water temperature monitoring employs a floating buoy to position a vertical axis for temperature measurement, and special purpose underwater cables and multiple deep water temperature sensors are suspended from the buoy, which automatically carry out acquisition of survey point water temperatures. The acquired monitoring data is sent to the telemetry equipment for data processing and storage.

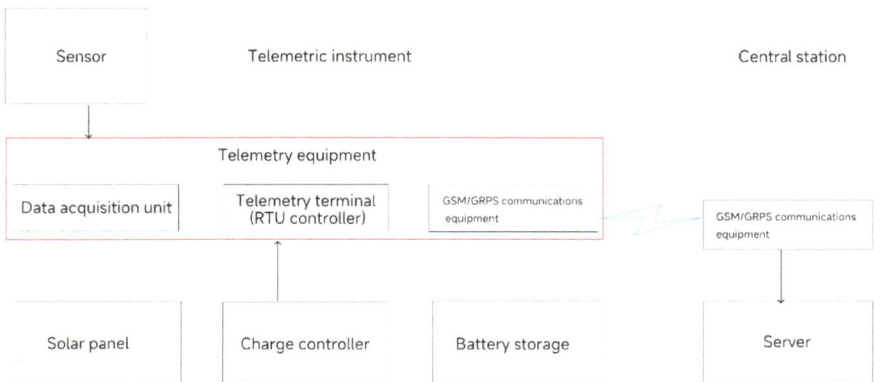

Fig. 8.1 Architecture of on-line remote telemetry station and central data receiving station

The telemetry station is powered by solar floating battery, and this power supply is not limited as to location, which is convenient for telemetry station deployment.

The water temperature central data receiving station consists of data receiving server, GSM communication module, uninterruptible power supply, AC charger, AC power lightning arrester, etc. Communication equipment sends data to the central station's receiver in GSM/GPRS format, and the data is stored in the central station's back-end server and also transmitted to the Dadu River big-data platform in real-time. The central station uses an external AC power supply and is equipped with uninterruptible power supply (UPS) to guard against incidents like power failure.

8.2.4.2   Data Library Management

After the on-line water temperature data acquisition is completed, the data is gathered and stored in the Dadu River big-data platform through the four steps of data extraction, cleaning, conversion, and transmission.

8.2.4.3   Real-Time Water Temperature Monitoring and Smart Control Platform

The real-time water temperature monitoring and smart control platform not only achieves GIS map display of cascade power station water temperature monitoring cross-sections and points, but through data processing and mining, it also achieves functions such as statistical analysis of historical discharge water temperature and reservoir area and down river water temperatures, and display of water temperature distribution mapping. Based on the changes of water temperature structure in the reservoir area and in water temperature of the discharge water body, the platform calculates and analyses the degree and pattern of impact of reservoir operation on river water temperature, and provides an assisted decision-making basis for further optimizing and supplementing the design of environmental protection measures to mitigate the impacts of reservoir water temperature, and for determining ecological water temperature dispatch solutions.

8.3 Case Application

8.3.1 Assessment of Fish Passage Effectiveness at Zhentouba (Level I) Hydropower Station

The first vertical slot fishway completed and put into use in the Dadu River watershed was the fishway at the Zhentouba (Level I) hydropower station. With a total length of 1,228.25 m, a top to bottom head difference of 34 m, and employing a vertical-slot cross-baffle channel body, this fishway was one of the world forerunners among high-head vertical slot fishways in design and construction.

The fishway of the Zhentouba (Level I) hydropower station is designed mainly for the passage of fish of the genus *Schizothorax* (*S. prenanti* and *S. davidi*), along with rare and endemic fish such as *Euchiloglanis davidi* and *Onychostoma daduensis*. After construction of the fishway, fish passage effectiveness was intelligently monitored and assessed using video observational analysis along with PIT tagging and tracking. A total of 453 fish were observed in the fishway by the video monitoring from April to July 2017, and 121 fish were observed in the fishway from March to September 2018. Observed fish included fish that live by crawling and clinging, such as *Euchiloglanis davidi*, *Liobagrus marginatus*, genus *Saurogobio*, genus *Glyptothorax*, and family *Gobiidae*, along with bottom-living fishes such as genus *Schizothorax*, family *Cyprinidae* (including *Carassius carassius and* genus *Onychostoma)*, and fishes that prefer to live in the middle and upper layers of the water body. Species passing through the fishway accounted for 52.83% of the total number of fish species below the dam. The dominant species below the dam can enter the fish passage smoothly, and the main fishway targets (*S. prenanti* and *S. davidi*) can pass through the fishway without issue. There was no obvious individual selectivity for the fishway. The lengths of passing fish ranged from 2.2 to 39 cm, with the fishway suitable for upstream passage of various sizes of fish. The actual fishway passage ratio was 71.7%. Continuous observation showed that the fishway at the Zhentouba (Level I) hydropower station was able to provide an upstream migration channel for many types of fish, and upstreaming showed clear diurnal differences, with more fish upstreaming at night and fewer fish upstreaming during the day.

In order to comprehensively assess fishway effectiveness and fish passage ratio at the Zhentouba (Level I) hydropower station, the Dadu River company further proposed an operation management solution that coordinates operation of the power station with fish passage results. This provided an intelligent result to reference for ecological dispatch and fishway operation of the Zhentouba (Level I) hydropower station, as well as for the research, design, operation, and management of fish passage facilities in the process of subsequent cascade development on the Dadu River.

8.3.2 Analysis of Real-Time Reservoir Water Temperature Monitoring at Pubugou Reservoir

The Pubugou hydropower station is the 17th cascade of the 22 cascade power stations planned for the Dadu River mainstem. It was developed in cascade as a key regulating reservoir for lower river. The dam height is 186 m, the total installed capacity is 3.3 million kW, the reservoir's normal water level is 850 m, the corresponding reservoir capacity is 50.64×10^8 m^3, the reservoir area is 84.14 km^2, and it is an incomplete annual regulation reservoir.

Pubugou reservoir is a typical high dam reservoir, with seasonal water temperature stratification. In order to comprehend and control reservoir and river water temperature in real-time, an on-line monitoring system for dam forebay and downstream water temperatures was built at Pubugou hydropower station, with dam forebay and downstream cross section water temperatures recorded by continuous observation at an interval of once an hour, with data retrieval on the hour each day. Water temperature data have unified formatting, rapid transmission, centralized receiving, as well as remote transmission and storage. So far, the big-data platform has recorded reservoir area and the down river water temperature data daily and hourly in real-time since the Pubugou hydropower station went into operation in 2013. The regular operation of the real-time water temperature monitoring and smart control platform has achieved real-time display of monitoring data, GIS map display of power station water temperature monitoring cross-sections and points, and display of discharge water temperature statistical charts and reservoir area water temperature distribution maps. Through application of a large quantity of water temperature monitoring data accumulated from long-term monitoring and the two-dimensional water temperature data model embedded in the smart control platform, the company has successfully analyzed water temperature structure for different time periods, water temperature stratification characteristics (thermocline layer changes, etc.), along with the pattern of dam forebay vertical water temperature change, the relationship between dam forebay water temperature distribution and temperature of the discharge water body, and the degree of latency in discharge water temperature, laying the foundation for achieving smart control of water temperature in the reservoir area.

8.3.3 Cascade Power Station Group Ecological Flow Control

The Dadu River Company operates eight cascade power stations, each set up with on-line facilities for monitoring of ecological flow discharge, real-time monitoring of power station discharge flow volume data, video monitoring of the actual conditions at the discharge outlet, real-time remote transmission of on-site information both to the local water resources supervision platform and to the Dadu River Company's self-built remote monitoring and management platform for ecological flow in the Dadu

River watershed, achieving sharing and comparative analysis between the enterprise's real-time monitoring data and the administrative authorities' real-time supervisory data. Utilizing the remote monitoring and management platform for ecological flow, an indicator system was set up for ecological flow discharge forecast and early warning, providing alerts via cell phone SMS to administrative personnel at all levels based on warning classification level. Based on this platform, ecological flow discharge can be appraised from the discharge compliance rate, equipment on-line rate, and automatic non-compliance alert rate, thus providing technical support for downstream aquatic ecological protection and ecological dispatch research.

Chapter 9
Intelligent Enterprise Management

9.1 Path Planning

The Dadu River Company has systematically planned the implementation of watershed hydropower enterprise intelligence, and has introduced a key path for business quantification, integration and centralization, unification of platforms, and smart cooperation.

9.1.1 Business Quantification

Lean enterprise management is achieved by scientifically setting standards and quantifying work tasks. That is, smart devices and IoT technology are used to acquire, transmit, and process all kinds of information data in real-time, and realize dynamic perception of various elements of the enterprise. By fully digitizing each enterprise business, the enterprise will gradually transform from the qualitative description and empirical management of the past, to data management, and the data itself speaking.

Business quantification is both a scientific management approach that makes business management measurable, and a basis for enterprises to "speak using data." Without this business quantification foundation, neither data analysis, nor any intelligence capability derived from it, would really be possible. To summarize, the Dadu River Company's approach is to fully quantify business into acquirable real-time data, covering not only the equipment in the hydropower production process, but also the production environment, human activities, and management processes. For example, by exploring the quantitative perception of employees' state of mind, the Dadu River Company also achieved the transformation of ideological and political work from passive perception of emotion, to active quantitative prediction and early warning management.

© The Author(s) 2025
Y. Tu, *Management of Hydropower Enterprises*, Water Resources Development
and Management, https://doi.org/10.1007/978-981-97-5584-4_9

9.1.2 Integration and Centralization

Through strategies such as overall planning, system consolidation, data centralization, and integrated operation, we eliminate the building of disparate business
systems, and the occurrence of fragmentation and data islands.

Integration and centralization is to solve left-over problems from the traditional
informatization stage, and to preclude their reappearance. As a large watershed-
scale hydropower enterprise, the Dadu River Company has many subordinate units
scattered in remote areas on the Dadu River watershed. Previously, almost all units
had their own server rooms, deployed their own databases, and developed their own
application software, such as production management systems, engineering management information systems, record keeping systems, etc. There was too much duplication in the enterprise's information infrastructure, resulting in a great waste of
resources. These systems were separated by great spatial distances, and by degrees
of expertise, resulting in issues of system isolation and data islands, which directly
led to information stagnation, duplication of operations, and other problems. Thus,
many of the developments in informatization not only did not increase efficiency,
but actually added to duplication of human labor instead. Starting in 2015, the Dadu
River Company set out to integrate the original company information infrastructure, establishing an enterprise-level big-data center to replace the original scattered
and independent computer rooms of each base-level unit, achieving the integration, centralization, and unified management of IT resources for the whole company
system. In the subsequent process of building intelligence, the Dadu River Company
insisted unremittingly on integration and centralization, primacy of planning, and
architecture-led practice. All new systems were required to be within the scope of
the unified planning blueprint, this blueprint including business architecture, system
architecture, data architecture, technology architecture, project roadmap, and even
a unified operations and maintenance strategy, so as to ensure that the intelligence
building process can be moved forward in systematic and orderly fashion.

9.1.3 Unified Platform

The data within various specialties are standardized and exchanged on a unified
application platform, and shared with each other in real-time, to provide support for
the sustained exploitation of the value of big-data.

The immediate purpose of the unified platform is to exploit and leverage data. After
business quantification, Dadu River Company further considered how to aggregate
the vast and dispersed data together, and put it to use for the business, and considered how to further create data-driven value to support enterprise management. If
integration and centralization is centralization at the hardware and software levels,
as well as centralization of the building process itself, then a unified platform is the
centralization and standardization of data. Before 2014, the information systems of

each business sector in Dadu River watershed had been developed relatively independently, and the data standards were not uniform, so there were barriers to data sharing, and low utilization of data assets. In 2015, utilizing the enterprise big-data center, and through unified planning, data governance, and centralized control, the enterprise big-data platform was established, which solved the lingering problems above.

9.1.4 Smart Cooperation

Through the specialty mining of big-data, we create various smart application models and algorithms, forming a "cloud brain" that automatically identifies risks, performs smart decision-making management, and cooperatively links multiple brains (including unit brains, specialty brains, and decision-making brains), and which carries out management of the enterprise, and achieves smart cooperation between people, systems, and equipment.

The core of watershed intelligent operation and management is the building of a digital enterprise on top of the physical enterprise. It is the enterprise brain, based on artificial intelligence smart algorithms, that carries out management, decision-making, and command of the digital enterprise. Operational unit brains have been built in the base-level units, and various specialty data centers (specialty brains) have been set up in company offices, and a decision-making command data center (decision-making brain), interconnected with the company-wide data center, has been assembled at the company decision-making level. This achieves vertical business connectivity and horizontal business cooperation, as well as the organic cooperation of people and equipment, systems, and algorithms. For example, at the unit brain level, in response to the shortcomings of the traditional point-to-point control approach between hydropower systems, along with highly complex networks and limited functionality, the Dadu River Company has not only developed a comprehensive data platform for power stations, realizing interconnectivity between systems, but building upon this, the Dadu River Company has also developed and deployed a multi-system smart linkage component for hydropower stations, realizing the active control and smart cooperation of multiple core systems, such as monitoring systems, ventilation, fire-fighting, video monitoring, gate control, etc.

9.2 Main Task

After clarifying the critical path, Dadu River Company established two key tasks. The first is to achieve the integration of physical and digital enterprises. The second is to achieve cooperation between people, and smart devices and systems.

9.2.1 Integration of Physical Enterprise and Digital Enterprise

The Dadu River Company was first to propose and put into practice the intelligent operation and management model, integrating physical and digital enterprises. Different from the digital twin concept found in German Industry 4.0 for production lines and equipment, what the Dadu River Company introduced was a digital twin of the entire enterprise management form. That is, while retaining the physical hierarchical organizational structure, and with focus on the digital transformation of core business and functional department specialty integration, change management system elements are gradually added, building a hierarchical management + data-driven "dual-track system" operational mechanism, progressively shifting the original management system towards increasing reliance on a data-driven management model. Operational mechanisms for intelligent operations and management are shown in Fig. 9.1.

From the operational mechanisms diagram, we can see that "people" have always been in the core position of enterprise intelligent operation and management. Between the physical enterprise and the digital enterprise there is a three step cyclic system prominently characterized by human–machine cooperation, allowing the inter-combination of hierarchical management and data-driven management for the whole Dadu River watershed, thus achieving intelligent operation and management for the watershed in the form of automatic anticipation, autonomous decision-making, and self-evolution.

The first-step of the cycle is to achieve the digital transformation of enterprises through people transforming the physical enterprise. The "people" (entrepreneurs and

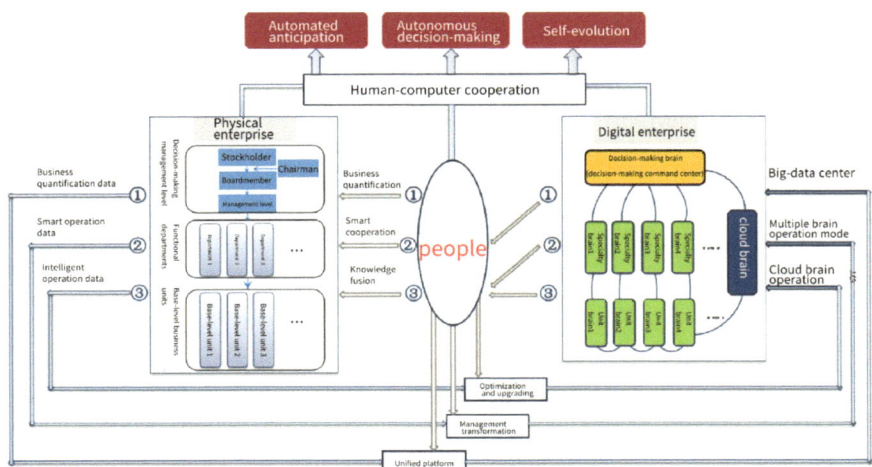

Fig. 9.1 Operational mechanisms for intelligent operations and management

employees) in the enterprise push forward the development of refinement and standardization within the enterprise, adopt advanced perception and transmission technologies, build big-perception and big-transmission systems, and realize the quantitative perception and integrated transmission of enterprise "things" and "human behavior." After quantification of perception, "people" carry out governance over the business data, and build a unified storage and operation management platform, thus realizing digital transition as characterized by the manifestation of enterprise big-data centers.

The second step of the cycle is to achieve application of enterprise smartness through people developing the digital enterprise. The "people" within the enterprise connect with the business needs of the enterprise, develop and mine the data in the enterprise's big-data center, and build various smart models (algorithms) within the digital enterprise, fashioning big-computing and big-analysis capabilities. The "people" evaluate and summarize the results produced operationally in the physical enterprise by the various digital enterprise models (algorithms). "People" constantly transform the physical enterprise production management system, so that it can be collaboratively integrated with various smart models (algorithms) within the digital enterprise, forming a smart operation capability, with multiple brains, such as decision-making brains, specialty brains, and unit brains, operating in a cooperative mode, these brains corresponding respectively to the decision-making management level, specialty department level, and unit base-level of the physical enterprise.

The third step of the cycle is to achieve intelligent operation of the enterprise through the integration of people and the smart enterprise. The "people" in the enterprise carry out the merger of their own innovation, creativity, and knowledge management achievements, with smart operation models (algorithms). They continuously optimize and upgrade operation models (algorithms) and management modes, so that the boundaries between the numerous brains within the digital enterprise, such as decision-making brains, specialty brains, and unit brains, gradually blur, and they merge into one, revealing an intelligent operation form characterized by cloud brain operation. The results of intelligent operation continue to fuse with human knowledge again and again in continuing cycles, further pushing the unceasing optimization and upgrading of the enterprise, achieving cycles of self-improvement and self-evolution of the enterprise.

9.2.2 Cooperation Between People and Intelligent Equipment and Systems

Cooperation between people and smart devices and systems is manifest at four levels.

The first is production line cooperation, that is, during the process of front-line production to achieve interaction and cooperation between people and smart devices. In the past, during production site operations, people and machines were separated and isolated from each other. On the one hand, machines can indeed substitute to carry out the repetitive labor of some people. Yet once machines experience something unforeseen, they must be immediately rescued or replaced by people, and at a site with continuous production, the timeliness of any remedy is very important. On the other hand, at a production site, people themselves are a highly variable element, so human operations being compliant to expectations is also very important. The smart terminal is an effective means to solve these problems. Through application of the smart inspection robot, machine inspection and manual inspection complement each other and act in cooperation, which not only improves the efficiency of inspection, but also greatly enhances the precision, validity, and flexibility of inspection work. Through application of smart helmets at the production site, the uncertainty and uncontrollability of human behavior is greatly reduced, achieving the efficient coordination of equipment operation and personnel behavior.

The second is production chain cooperation, that is, optimization and cooperation in process, division of labor, and organization. Whether digitalization, networking, smartness, or intelligence, the purpose of its transition is to achieve organizational efficiency, rather than the transition of a constituent business part or a simple department. In order to achieve this goal, the determined and concerted efforts of the entire organization are required, and the medium for this labor is process, division of labor, and even the entire organizational structure. Therefore, once having attained cooperation on the production line, the Dadu River Company further adjusted the division of labor along the production chain, for example, eliminating the central control rooms of the individual power plants through the construction of a command center for the whole watershed, and eliminating the hydraulic engineering teams of each power plant through the establishment of the watershed reservoir and dam operation safety center. These actions led to the production process being reduced by about 120 and 150 people, respectively, and the production chain being optimized.

The third is the management system cooperation, that is, the requirement for continuous transformation and upgrading of enterprise management. The wide application of digital technology, represented by big-data and artificial intelligence, will inevitably bring new ways of production and of living with the three-way integration of people, machines, and things. For enterprises, this means new ways of organization, that is, new management systems. By merging company office departments and transforming company office functionality, establishing a big-data service company, and setting up innovation workstations, the Dadu River Company has liberated its employees from operational positions having difficult environmental conditions, allowing more employees to engage in creative work, and the whole organization's capability for innovation has been significantly improved. Currently, 12% of employees have transferred to digital-related work, and organization and management have taken on a whole new look.

The fourth is ecological chain cooperation, that is, the requirement for enterprises to go from closed enterprise boundaries to an open industrial ecology. This is an inevitable requirement for the development of the digital economy era, based on the flow of data breaking industrial barriers and achieving both industrial digitization and digital industrialization. The Dadu River Company set up a technology platform company and built an industry-oriented data service platform, creating specialty service capabilities, providing resource and capability sharing services for the industry, and forming an initial ecology of sharing, integration, and cooperation in geological hazard warning, hydrometeorological forecasting, power market big-data analysis, equipment health status assessment, smart equipment, emergency rescue, and other areas.

9.3 Exploration and Practice

9.3.1 Traditional Management Period

The Dadu River Company was established in the year 2000. Starting with the Gongzui and Tongjiezi power stations "parent" projects built in succession, the company has carried out comprehensive development of hydropower resources in the Dadu River watershed, providing energy assurance for the economic development of Sichuan Province. The traditional management period of the company can be divided into 3 stages:

9.3.1.1 Construction and Operations of Parent Power Plants Stage (1964-2000)

The earliest power station in the Dadu River water was the Gongzui power station, which started construction in March, 1966. The first generator set produced electricity in December 1971, and all generator sets were commissioned by 1978, this station thus becoming the earliest "parent" power station in the Dadu River watershed.

In order to promote full socioeconomic development, deepen reform and development, further drive the rapid development of the electric power industry, and ease the serious shortage of electricity in Sichuan, the government took the lead in assembling funds to start the construction of Tongjiezi hydropower station. Construction work formally started in 1985, the first generator set producing electricity in December 1992, and the project fully completed by December 1994.

In order to implement China's "Great Western Development" and "West-to-east power transmission" strategies, and to develop the pillar industry of hydropower in Sichuan, the State Grid Sichuan Electric Power Company, and the Guodian Power Development Company Limited decided to reorganize the assets of the Gongzui hydropower plant, making Gongzui hydropower plant one of the hydropower plants at the forefront of power system reform, and representing an important step towards the reform goals of dividing power plants from power networks, and moving bidding on-line. On November 18, 2000, Dadu River Company was officially established.

9.3.1.2   Watershed Rolling Development Start-up Stage (2001–2010)

The first project for the newly established Dadu River Company was construction of the Pubugou hydropower station, this starting point opening the curtain to rolling development in the watershed. The company has unswervingly implemented watershed, cascade, rolling, and comprehensive development, and unswervingly advanced the principle of the central position of watershed development, which is now well established, and the three fronts of power production, basic infrastructure, and integrated development, which have maintained a good momentum of sustained, rapid, cooperative, and healthy development.

On December 27, 2001, the engineering construction department at Pubugou hydropower station was officially established, marking the acceleration of rolling development of the watershed. The Pubugou hydropower station was a construction project started under the national "Tenth Five-Year Plan," and also a key project in China's western development. At the end of 2009, the first two generators sets were put into operation, and by the next year all generator sets had been fully commissioned.

In May 2005, in accordance with deepening work deployment reform, the original general hydropower plant at Gongzui, which managed both itself and the power plant at Tongjiezi, was reorganized into three separate units: A new general power plant, a watershed servicing and installation branch company, and an industrial company, forming a new mode of power production management.

On December 26, 2008, under the preparatory principle of unified planning and step-by-step implementation, the centralized control center for cascade power stations was put into operation, so as to build a vertically connected and horizontally integrated unitary platform for enterprise-level production and operation, and realize watershed unified dispatch. The commission of the centralized control center marked the beginning of the era of unified dispatch in the Dadu River watershed.

9.3.1.3   Accelerated Watershed Development Stage (2011–2013)

In order to meet the development needs of the new situation, and centered around the enterprise transition goals of shifting from a focus on mainly developing hydropower projects, to becoming a management and operation enterprise, from mainstem development and operation, to all-round hydropower development and investment, from simplistically developing hydropower, to becoming a hydropower-centered comprehensive enterprise, from a specialized company, to an open and responsible social enterprise, and building a first-class international watershed hydropower company that is efficient, pioneering, innovative and harmonious, the Dadu River Company accelerated the development of the mainstem of the Dadu River, and at the same time undertook the construction of Shenxigou, Dagangshan, Houziyan, Zhentouba (Level 1), and Shaping (Level II) power stations, and the company increased efforts in mergers and acquisitions of small and medium-sized hydropower, completed the acquisition of the Geshizha Company, built the Jiniu power station, and accelerated the development of solar, wind power, biomass power generation, as well as non-electric general industries.

By 2014, the scale had continued to expand, with the installed capacity on the Dadu River watershed increasing from 1.3 million kW in 2000, to 5.920 million kW, having quadrupled. The total number of employees had also increased from 1,519 in 2000, to 2,124 in 2014, an increase of nearly 40%. The organizational structure had become more sound, and the watershed-based and professional management mechanism had gradually matured, all of which laid the foundation for the explosive growth on the Dadu River watershed after 2014.

9.3.2 Intelligent Management Exploration and Practice Stage

The Dadu River Company's intelligent management transformation began in 2014. With the gradual maturing of new technologies such as cloud computing, big-data, IoT, mobile internet, artificial intelligence, and their beginning rise within enterprises, the Dadu River Company began to realize that the digital technology revolution marked by these new technologies was coming to life. At the macro level this would change industrial and economic structures, and greatly enhance social productive forces. At the micro level, enterprise management modes would change, the efficiency of organizational operation would improve, and the new foundational infrastructure composed of these technologies would occupy an unassailable position in the operation and management processes of the enterprise.

Given this context, the internal and external environment of enterprises in the new digital era would undergo revolutionary changes. The Dadu River Company actively embraced change, and actively initiated transition, starting with cultural transition, laying the foundation for full personnel participation and overall transformation. From 2014 to 2020, Dadu River Company management reform went through three stages, as shown in Fig. 9.2.

9.3.2.1 Cultural Transition, Business Perception Quantification Perception Stage

The period from 2014 to 2016 was the first stage of Dadu River management transformation. During this stage, the Dadu River Company carried out a transition of culture, with the goal of business quantification perception, and a push for changes to personnel positions. This process focused particularly on strategic leading and improvement of leadership, cultivation of innovation culture, and participation by all personnel.

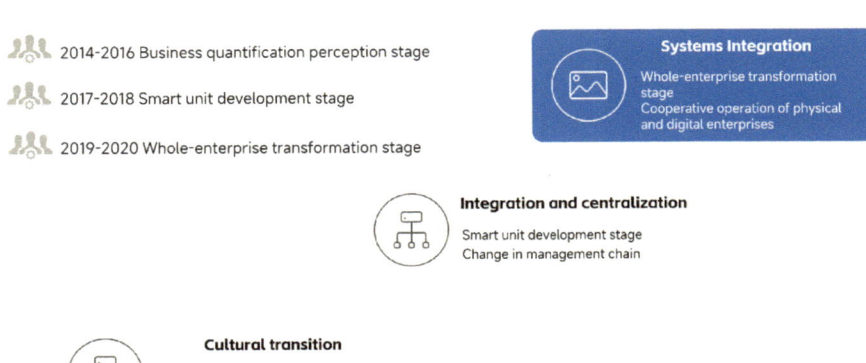

Fig. 9.2   The stages of management transformation in the Dadu River Company

Emphasis on strategic leading and improvement of leadership: Strategy and leadership are like the foundation and roof of a building, with strategy covering the overall direction of the design of management transformation, and leadership being the foundation of all transition. The Dadu River Company's strategy of building an intelligent enterprise started in 2014. During that year, the Dadu River Company no longer confined its thinking on management change to the hydropower industry, but rather gathered experts from different industries and invited academics from different fields, representing the top level in China, to jointly participate in the study of intelligent enterprise theory.

With the deepening of research into intelligent enterprise theory, a detailed vision for intelligent enterprise emerged, and the transition path of business quantification, integration and centralization, unified platform, and intelligent cooperation was proposed, thus giving a clear direction for management transformation. This was followed by the establishment of the leading group for developing the intelligent enterprise, and thus management change, presided over by the hand of enterprise, was formally kicked off.

Emphasis on the two pillars of enterprise innovation and enterprise culture: Supporting strategy at the two pillars of corporate innovation and culture. Without these two pillars, it is impossible to bolster the implementation of management transformation strategy, and any transformation will be reduced to mediocrity in the shock of continuous unstopping readjustment and rebalance. The Dadu River Company was deeply aware that neither innovation nor culture can be shaped in one day, and a long-term battle against psychological inertia, mindset inertia, and behavioral inertia of personnel was required. For this reason, Dadu River Company established intelligent enterprise discussion salons, and youth innovation workstations, which became two powerful tools for forging and spreading innovation culture.

The youth innovation workstations and intelligent enterprise salons created a completely new innovation culture network for Dadu River Company. The youth innovation workstations became a cradle to inspire innovation, allowing a number of innovations spearheaded by young employees to emerge within the company. The intelligent enterprise salons became a sharp blade to disrupt the traditional stability, and through the weekly salon, all kinds of innovative views could come together and collide in one place, such that employees spontaneously started to talk about intelligent enterprise, and innovation culture started to spread.

In addition, the traditional way of management in the Dadu River Company began to change. An enterprise's routine way of operating is the most ordinary way of expressing enterprise culture, and the ordinary way the enterprise handles business is a representation of the behavior pattern of enterprise personnel. Based on the penetration of innovation culture and the transmission of the unified vision of the intelligent enterprise, the Dadu River Company stimulated the enthusiasm of the employees to contribute to the intelligent enterprise transformation of management, in consequence influencing the daily work style and, in the end, even producing an impact on the way of running the enterprise. The focus of employees began to shift from transaction processing to focusing on business itself, paying more attention to how to perceive and quantify business, and then moving towards an integrated, centralized, and unified platform, and an intelligent cooperation path, which became a common manifesto for all employees.

Emphasis on actions that lead to participation of all personnel: The goal to be achieved at this stage, whether in putting strategy or culture into practice, was to achieve whole-business quantitative perception. Innovation and culture are the contextualization and medium, and all personnel start from their own work positions, and then seek the help of digital technology on their own initiative. Through the coding for each business object, and the design of their individual state quantities, along with risk point acquisition, business becomes distinct figures that can be measured, so that big-data, artificial intelligence, and other technologies can come into use. But also, the work position's original role has the potential to be divided in two, half going to the machine, and half going to the person.

The work center of gravity for personnel progressively experienced change. Eventually, work positions started to undergo adjustment, and the organizational structure gradually started to change, implementation of management transformation naturally followed.

9.3.2.2 Integration, Centralization, and Smart Unit Development Stage

In 2017 and 2018, during the second stage of management transformation, the Dadu River Company carried out integration and centralization transitions, focusing on the development of smart units, and promoting changes in the business chain. The most important result of this stage was the emergence of a new enterprise management "brain" centered on data.

In the company's management transformation, the establishing of smart units was the process of shifting from the "point" of the work position, towards the "line" of the business chain, and the "plane" of the management chain. At this stage, the reform of Dadu River Company headquarters and of institutions at all levels on the watershed was in full motion, and the traditional stable pyramidal organizational structure had gradually evolved into a flexible organizational structure of "one hub, multiple centers, and four units." Twenty-one data centers and four major units were established, for a total of 25 smart units.

The process of establishing these 25 smart units was a process of deeply integrating business and digital technology. Each smart unit breaks the boundaries of traditional functional department compartmentalization, and reorganization is in the form of data, and accordingly, people are also reorganized around data, making enterprise business processes interweave with data processes, thus forming a bundle of completely new data-driven business, management, and decision-making capabilities.

Establish 21 data centers for full empowerment: At the level of decision-making command, a decision-making command center was established to effectively perceive the major risks of the company, primarily by gathering data from the whole watershed, realizing the whole process control of major focus areas and important production operation indicators, and the dynamic analysis of anomaly causes, and establishing and fortifying the company's emergency response smart decision-making command support system for geological threats, flood and high-water control, incident clusters, and other emergencies, so as to finally realize smart control of major risks, control of major business processes, and smart support of major decision-making.

At the specialty department level, 20 specialty data centers were established, which can be further divided into business chains, management chains, and data services.

In the business chain, data capitalization was carried out on the knowledge accumulated over a long term, using digital technology as represented by big-data and artificial intelligence. Eight data centers were established to enhance specialty control capabilities (including project control data center, geological hazard prediction and early warning center, watershed ecological environment monitoring center, equipment control data center, dam safety control center, hydrometeorological data center, power market data center, and safety control data center). Even external empowerment, based on advanced smart algorithms, was done to raise professional levels in related industries. The establishment of these eight data centers opened up the control between power production operation and the company's headquarters. Production site data scattered across the Dadu River watershed were able to be integrated and centralized into the eight data centers, realizing the vertical connectivity of hydropower production operation and management. The physical power plants previously scattered across the watershed, and their representative production capacities, were integrated and centralized into the eight data centers, which at the same time meant that the organizational personnel, and their representative specialist capacities, scattered across different management chains, could break the traditional organizational and hierarchical constraints and be integrated and centralized under the command of the data centers, so that the entire production operation process could achieve data-driven whole-watershed resource optimization.

In the management chain, the traditional "people, finances, and things," along with public administration itself, are designed to serve the enterprise, but over the long-term running of the enterprise, rigidity easily creeps in, and complex processes and reporting bring additional burdens to personnel. Dadu River Company has returned management to the essence of business empowerment, establishing eleven specialist data centers (including intelligent Communist Party development center, human resources data center, material control center, financial shared service center, procurement and contract data center, discipline inspection data center, audit data center, legal affairs center, vehicle control center, digital buildings center, and digital record keeping center), replacing repetitive work with smart algorithms and processes, and achieving better enterprise services based on data analysis. The eleven specialty data centers on the management chain have become eleven capitalization service units for the enterprise, and together with the eight capital creation units on the business chain, they exhibited initial smart cooperation. These eleven specialty data centers shoulder the responsibility of achieving centralized control of functions, breaking through information exchange barriers, optimizing resource allocation, etc., and are the assistants and supporters of enterprise operation and management. This is the integration and centralization of enterprise management capabilities. Although the management chain does not produce direct economic benefits for the enterprise, but as the processes scattered between traditional functional departments were unified, the data centers made is so that traditional functional management saw transformation from management towards service, and since the indirect value chain can maximally achieve empowerment of the direct value chain, the value of management itself was further enhanced.

In terms of data services, one big-data center was established to provide unified data services for all the data centers, above.

With the construction of the 21 specialist centers, the personnel roles and division of labor gradually transformed to specialist data maintenance and application, as well as innovative product development. The management process was optimized. Specialist and intensive management, and centralized control of key domains and key constituent parts, changed the previous traditional mode of specialist plans being submitted from the base level to the company offices, and being reviewed gradually level by level. This improved the efficiency and management level.

Establish four major units to push capital creation: Based on Dadu River Company business characteristics, more than 20 units were reorganized into the four major units of intelligent engineering, intelligent power plant, intelligent dispatch, and intelligent servicing. By deeply integrating advanced technologies such as IoT, mobile internet, and artificial intelligence, multi-system linkage and comprehensive perception was accomplished.

The construction of the four major units brought significant changes to the company's production operation mode. Organizational structure was streamlined and compressed, with implementation in orderly steps according to categories, priority given for consolidation of specialties such as safety monitoring, power generation dispatch, and economic operation, along with computer networking, based on the degree of centralization of the specialty. Base-level power plants was no longer set up departments and teams for servicing, dispatch, hydraulic observation, and network information, thus realizing unified management without hierarchy. Additionally, production personnel could achieve quality and efficiency improvements without increasing head counts. Based on the application of smart equipment, repetitive, high-risk, and low-technology tasks involving daily equipment rounds, regular periodic safety monitoring, etc., were gradually handed over for robots to complete, and work involving economic operation solution optimization, cascade generation load adjustment, and generator set health status assessment, etc., was accurately calculated using big-data, and handed over to the smart system for decision-making. The emergence of smart cooperation allowed the value of both machines and people come into full play.

9.3.2.3   Systems Integration and Whole-Enterprise Transformation Stage

In 2019 and 2020, during the third stage of management transformation, the company carried out systems integration transition, with whole-enterprise transformation as the core, pushing forward the emergence of the digital enterprise. The most important result of this phase was the cooperative operation of the total physical and digital enterprises. The whole appearance of the enterprise shifted towards a human–machine cooperation mode.

With the establishment of smart units, the organizational model of "one hub, multiple centers, and four units" gradually matured, and the Dadu River Company started to develop smart cooperation at a higher system level, evolving from a transition of "plane," to a transition of "space." A specialized big-data company was launched, meaning that after the Dadu River Company achieved enterprise vertical connectivity and horizontal cooperation, each smart unit became a step further integrated, and through the big-data company, each smart unit was provided with data technology as the core of a unified service. The basic infrastructure of the digital enterprise was solidified, and "data-driven" as a principle penetrated the whole enterprise. This is a systematic transformation of the enterprise, which helped bring about the birth of a completely new model of intelligent operation on the Dadu River watershed.

The big-data company performs centralized operations management of the company's hydropower whole industry chain data resources, providing data governance, data mining, data products, data operations, and other consolidated data services for each specialty data center, building an enterprise-level data resource pool, realizing multi-source data integration, continuously accumulating the various specialty data center assets, and safeguarding the building of digital enterprise in the Dadu River watershed. With the dynamic evolution in development of the intelligent enterprise, the big-data company became fully responsible for internal data operations of the Dadu River Company and its subordinate units, responsible for providing R&D, services, maintenance, and assurance for data centers such as specialty brains, unit brains, and decision-making brains, providing forceful information support for business management. Externally, the big-data company became responsible for summarizing the achievement of Dadu River Company intelligent enterprise development, and carrying out conversion of achievements, and export of technology.

The establishment of the big-data company represented a shift from management transformation to systems transformation. At this stage, the focus of the overall transformation was to open up data connectivity. The digital enterprise started to emerge, and run in cooperation with the physical enterprise.

On the whole, the Dadu River Company is a traditional hierarchical physical enterprise, on which a corresponding non-hierarchical digital enterprise is built. The smart and cooperative merging of the hierarchical physical enterprise and the non-hierarchical digital enterprise allows the hydropower enterprise to take on an intelligent operation state of automated anticipation, autonomous decision-making, and self-evolution. Dadu River Company intelligent operation model is shown in Fig. 9.3.

The digital enterprise mainly focuses on creating decision-making brains at the strategic decision-making level, specialty brains at the business management level, and a unit brains at the production operation level, opening up the whole management chain through real-time data interchange between these three brains, building a cooperative operation mode for the three brains, promoting efficient operation of the digital enterprise, realizing data-driven and intelligent cooperation, and providing decision-making support for physical enterprise decision-making management.

In between the physical and digital enterprises, smarter production, more flexible organization, and more scientific management are promoted, with a focus surrounding deep integration of informatization, industrialization, and management

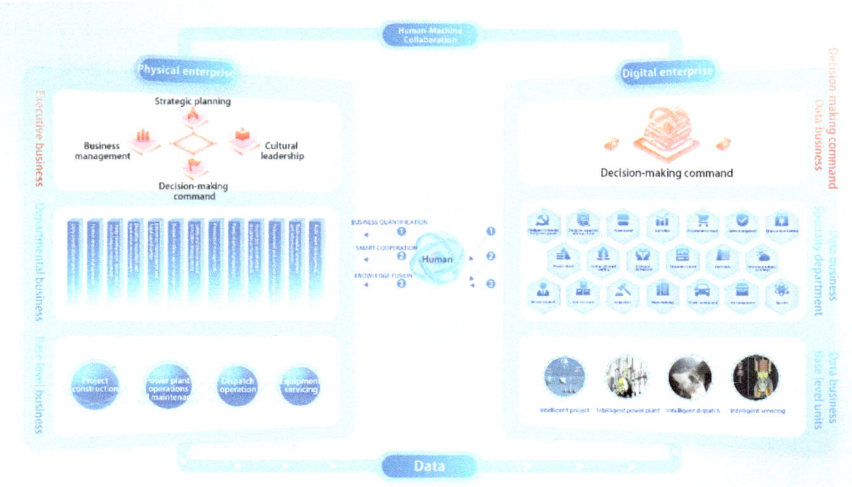

Fig. 9.3 Dadu River Company intelligent operation model

modernization, so that the enterprise can achieve intelligent operation of auto-mated risk identification, smart decision-making management, and autonomous remediation escalation.

9.4   Early Results

From 2014 to 2020, management transformation in the Dadu River Company has achieved substantial results. By the end of 2014, five power stations had been put into operation in the Dadu River watershed, with an installed capacity of 5.87 million kW. By the end of 2019, nine power stations had been put into operation, with an installed capacity of 11.34 million kW, doubling the number of power stations and installed capacity. At the same time, the total number of employees in the company had been reduced from 2,498 to 2,148, a personnel reduction of 14%, and total labor productivity had increased by 75%, as shown in Fig. 9.4.

Through three stages of management transformation, the advantages of intelligent operation have gradually become clear.

9.4.1   Shift of Management Mode from a Hierarchical System to a Center System

Centered around the intelligent enterprise top-level design architecture of "one hub, multiple centers, and four units," the company deepened management system trans-formation, and advanced the reinvention of process mechanisms. Optimizing and consolidating the company offices and associated base-level specialty institutions,

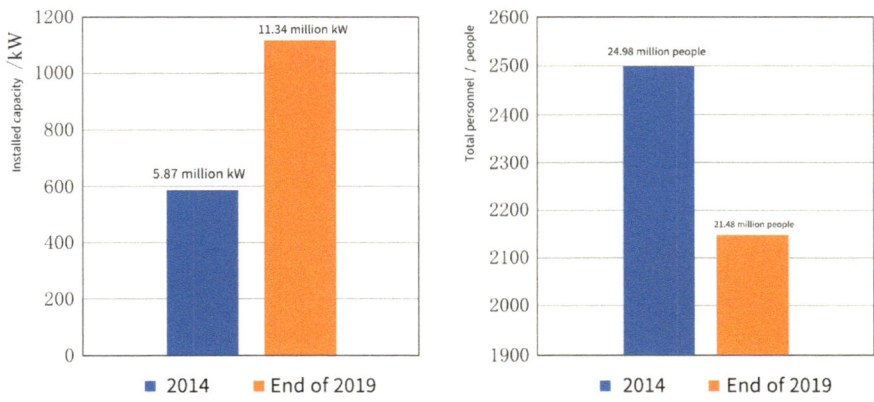

Fig. 9.4 Statistics for installed capacity and total personnel numbers on the Dadu River watershed

21 specialty data centers were established at the company headquarters. At projects in the midst of construction at Shuangjiangkou, Jinchuan, and other locations, the development of intelligent projects was promoted following the center system framework, forming a more specialized and flat management model, breaking the traditional management barriers between levels and departments, and promoting the optimization of human resources and streamlining of management institutions.

9.4.2 Shift of Production Management from Manual to Smart

The innovative application of a number of new technologies such as adaptive unmanned roller-compacting, smart mixing of core wall materials, and project dynamic design has improved the level of smart management of the construction site. Related research achievements have been appraised by experts as reaching internationally advanced levels. Technological means such as UAVs in the air, robots under the water, and smart USVs on the water, are employed to carry out reservoir area monitoring of sediment siltation and high side slope stability. The company has vigorously promoted innovative technologies such as self-developed smart robots, multi-system linked platforms, and 3D digital plant buildings, which reduce various types of repetitive, high-risk, and low-technology work. The company has established a modern production mode of "staff-less shifts and few people on duty," scientifically formulating economic operation solutions, smartly matching up power generation load adjustments, monitoring the health status of generator sets in real-time, realizing the smart autonomous operation of hydropower projects with 10 million kW of installed capacity in the Dadu River watershed.

9.4.3 Shift of Decision-Making Command from Experience-Based to Data-Based

The research and development of economic dispatch control (one-button dispatch) technology for watershed power stations transforms the previous mode of provincial grid dispatch issuing load targets to a single power station and a single generator set, into a mode of issuing an overall load target to multiple power station groups. This eliminated the workload of more 30,000 load adjustments a year, and zero manual intervention for load adjustment was achieved at three stations for a whole year. Through the application of advanced research achievements such as quantitative precipitation forecasting, flood water is utilized as a resource, smart dispatch decision-making support, and economic dispatch control, the power generation efficiency of the enterprise has been greatly improved. The company has established an on-line equipment condition detection platform, carried out equipment operation big-data analysis, monitored and diagnosed generator sets in real-time on-line, advance anticipation of equipment operation health status, deployed timely servicing solutions and strategies, and increased the equipment equivalent availability factor by about 10%.

9.4.4   Shift from Passive to Anticipative Risk Control

In equipment management, an intelligent servicing system is applied to realize objective assessment of equipment health status, and to anticipate equipment health change trends, shifting from traditional planned servicing, to intelligent servicing prediction and anticipation. In forecast and dispatch, we make full use of high-precision hydrometeorological monitoring and reporting means to quantitively analyze watershed flood and rainfall information, and automatically project the flood water regulation process, to provide decision-making substantiation for flood regulation. In geological hazard control, the company makes full use of geological hazard prediction and early warning technology, and has repeatedly achieved geological hazard prediction and early warning in the Dadu River watershed.

9.4.5   Shift from a Production to an Innovation Workforce

The company actively cultivates teams made up of innovative personnel having capabilities in multiple specialties, and also fortifies innovation entities, and promotes the shift of production personnel focus from conventional shift work operations, to risk control, emergency handling, big-data application, innovative product development, etc. Staying centered around on the core business of hydropower, the company built six industry tracks for enterprise digital transition, edge computing, smart applications, industrial internet, energy conservation and environmental protection, and safety management big-data, and successfully incubated innovative products such as smart inspection robots, smart helmets, and smart keys to create anticipated economic benefit growth points. The Dadu River Company was established in 2000, and started the exploration and practice in intelligent operation and management in 2014. In those intervening five years, the company obtained 350 instances of intellectual property rights, which is 5.6 times of the previous 15 years. The company organized the compilation or participated in compilation of 39 national and industry standards, including wholly new breakthrough standards. The company won 65 national and provincial science and technology progress awards, which is 1.1 times more than the sum of the previous 15 years put together.

9.5 Looking Forward 10 Years

The completion of the overall smart cooperation model for the physical and digital enterprises, means that Dadu River Company has essentially completed the building of a new kind of enterprise management capability that is based on a new round of digital technology infrastructure, and the next step will be to enter a completely new stage of business empowerment and creation. With the goal of continuously improving "human self-worth realization," the Dadu River Company has formed a plan and is looking forward at the next ten years, from 2021 to 2030.

9.5.1 Stage 1: Reduce One Third of the Staff of the Site Management Units and Production Operation Personnel Within Three Years

The focus of this stage is the establishment of intelligent power plants. The Dadu River Company initiated planning related to intelligent power plants very early, and took the lead in compiling industry standards, such as the "Technical Guidelines for Intelligent Hydropower Plants," taking a leading position in the industry in the development of intelligent power plants. With this foundation, the Dadu River Company has proposed the concept of the smart autonomous, unmanned power plant, the fundamental purpose of which is to assure the efficient operation of the power plant and free operating unit personnel from the harsh conditions of the base-level production environment. The Dadu River Company already has mature and favorable conditions for this. For example, the capability for smart autonomous operation of power plant equipment and systems has taken initial shape, and manual operation tasks have been greatly reduced. Power plant perception capability has been enhanced, and capability for all-round three-dimensional space-aerial-terrestrial perception has largely been built. Tens-of-thousands of technological eyes are assisting operation personnel to provide monitoring and anticipation capability. Cognitive capability for power plant operation has gradually been established through data analysis capability, and the overall production management mode has shifted from careful safeguards to risk response.

Based on these achievements, operation units have started to realign towards smart autonomous, unmanned power plant operation mode. At present, there are already power plants that have rearranged the conventional operations and maintenance team into two flexible departments, to respectively assume the major responsibilities of enhancement of smart autonomous capabilities, and enhancement of unmanned power plant risk response capacity. With the improvement of these two major capabilities, the company can further achieve the goal of reducing the operation staff by one-third within three years, and further merging the functional departments of the operation units, fully promoting the "remote cloud office," and setting up on-site divisions responsible for work such as on-site emergency handling. Daily work can be carried out at the company headquarters in the city of Chengdu, Sichuan Province, with personnel visiting on-site to work as needed, which will greatly enhance the happiness of front-line staff.

9.5.2 Stage 2: Reduce Site Management Units and Production Operation staff by Half Within Five Years

With the improvement of the smart and autonomous capability of power plants and the risk response capability of unmanned power plants, based on the distribution of power plants in the Dadu River watershed, power plants can be further reorganized into regional power plants groupings on the upper, middle, and lower river regions, to realize the "1 + 3" regional power plant operation mode:

"*1*": One centralized and unified production dispatch command center located at company headquarters, responsible for operation monitoring, production dispatch, and economic operation of the whole watershed, achieving flexible and mobile arrangement of equipment servicing strategy and rapid emergency command under dynamic working conditions.

"*3*": Three regional power plant management centers, which respectively manage power stations in the upper, middle, and lower watershed regions. Each power station would have a minimum scale of duty personnel, with dynamic rotating rounds within the scope of the regional power plant management centers. Each of the three regional power plant management centers would be set up in the best locations, where all power stations in the regions each respectively administers can be reached by fast transport, and special operation detachments for servicing and emergency repair would be set up, to realize flexible deployment and dynamic support of personnel in each station in the region.

This stage will fundamentally change the conventional production operation and management mode of fixed duty and regular periodic servicing, and realize the maximum utilization of human resources under conditions of limited personnel. Based on the merging of specialties and the mobile response of the equipment management special detachment, the institutions and personnel of the operation unit can be greatly reduced, and the personnel will be liberated from front-line work and can turn to work on analysis, management, and decision-making.

9.5.3 Stage 3: Eliminate Site Management Units and Production Operation Staff Within Ten Years

With the development of digital technology and the full cooperative operation of physical and digital enterprises, the Dadu River Company will gradually realize the advanced goal of enterprise self-correction and improvement, which is mainly manifested as follows: All the deterministic work of the enterprise will be completed in the digital enterprise, and only under support of the digital enterprise will people need to deal with the risks within the physical enterprise. Management of the enterprise will be left as much as possible to the smart system within the digital enterprise, so that the hierarchy within the physical enterprise can be completely eliminated.

In the base-level units, the power stations will achieve smart autonomous, unmanned power station operation, thanks to future technological developments. Servicing and emergency operation detachments made up of people will not need to be stationed in the vicinity of the power stations, and rapid handling of site emergencies will be rapidly carried out based on technology, the sites only needing to stock the appropriate material resources. Under such circumstances, there will be no concept of the company offices or the base-level in the Dadu River watershed, and everyone will be able to study how to empower the enterprise and themselves based on the latest digital technology from the comfortable environment of Chengdu. The power stations will be running deep in the mountains behind closed doors. Occasionally people will go on site, not to operate the machines, but to seek out more inspiration for their work, while enjoying the sun, air, and beauty of nature, as people pursue the realization of their self-worth in creative work.

Postscript

Intelligent hydropower operation and management is a new type of operation and management paradigm based on the deep integration of big-data driven smart models with mechanism models based on human empirical knowledge. This is a required pathway for hydropower enterprises to apply new technologies such as cloud computing, big data, IoT, mobile internet, and artificial intelligence to achieve enterprise transition and improvement. I hope through this book to explore this new paradigm together with friends from all fields.

From book proposal and conception, to preparation, discussion, revision, and translation, has taken nearly four years, with the manuscript going through repeated revisions before final fruition. During the compilation process, I received encouragement and guidance from Prof. Biswas and Dr. Tortajada, as well as help from our hydropower colleagues such as CHEN Gang, ZHENG Xiaohua, TAO Chunhua, YANG Gengxin, HUANG Lingmei, LIU Jing, CHEN Zaini, CHEN Bangfu, WEN Weijun, MA Yue, HU Hanyin, ZHONG Qingxiang, HUANG Xiang, WANG Fuzhi, LIU Yu, LI Guanghua, SHANG Chunhai, and others. The book also benefitted from the careful efforts of the book's translator, Dr. Patrick Lucas. My thanks to you all!

Although the content and viewpoints of this book have been discussed and revised in depth numerous times, there are undoubtedly still many shortcomings and deficiencies, or even errors, given the wide scope of the content involved, the high specialization, and the difficulty of research, along with limitations of theory, vision, and perspective of the writer, so I am extremely grateful for the patience of readers, and look forward to input and corrections from them.

References

Chen J, Tao C, Ma G, Chen S, Zhao Y, Wang J (2020) Jiyu shuju wajue yu zhichi xiangliangji de xianhuo shichang chuqingjia yuce fangfa [Spot market clearing price prediction method based on data mining and support vector machine]. Dianwang yu Qingjie Nengyuan 36(10):14–19+27

Cui P (2020) Jiyu GNSS ji celiang jiqiren de daba anquan jiance yanjiu—Yi Zhentouba shuidianzhan wei li [Research on dam safety monitoring based on GNSS and measuring robots: a case study of the Zhentouba power station]. Renmin Changjiang 51(S1):132–134+148

Geng Q, Ma Y, Feng Z (2017) Xinxingshi xia shuili fadian qiye jianshe zhihui jianxiu de sikao [Thoughts on developing intelligent servicing for hydropower enterprises under the new situation]. Shuidian yu Xinnengyuan (12):44–45+64

Geng Q, Zhang H, Feng Z (2016) Shuilun fadian jizu zhihui jianxiu jianshe tanxi [Analysis on development of intelligent servicing for turbine-generator sets]. Shuidian yu Xinnengyuan (09):8–12

Ju S, Huang H, Feng T (2019) Daba anquan jiance xinxi jizhong jicheng yu zaixian jiankong guanli shijian [Practice in centralization and integration of dam safety monitoring information, and online monitoring and management]. Daba yu Anquan (02):42–46

Li C, Ke H (2016) Pubugou dianzhan daba waibu bianxing jiance zidonghua xitong sheji ji yingyong [Design and application of automated system for external deformation monitoring for Pubugou power station]. Shuidian Yu Xinnengyuan 10 8 12

Li L, Hou Y, Zheng J, Liu R (2019) Jiyu quanshengming de shuidianzhan zhihui jianxiu chuangxin shijian [Innovative practice in full life cycle-based hydropower station intelligent servicing]. Shuidianzhan Jidian Jishu 42(12):31–34+101

Li X, Zhong Q (2020) Liuyu tiji shuidianzhan fuhe zhineng tiaokong moshi yangjiu [Study on watershed cascade hydropower station smart regulation mode]. Sichuan Shuili Fadian 39(06):130–135

Liu H, Liu F, Wu L (2019) Shuidianzhan shineng xunjian xitong sheji. [Hydropower station smart inspection system design]. Shuidianzhan Jidian Jishu 42(12):24–27

Tang Y, Liu H, Zhang L, Liu X (2019) Pubugou dianchang zhihui shuidian jianshe shijian [The practice of intelligent hydropower construction at the Pubugou power plant]. Reli Fadian 48(09):156–160

Tao C, Ma G, Tu Y, Chen G, Liu J, Tang M (2007) Yizhong dianzhan changnei jingji yunxing suanfa [An in-plant economic operation algorithm of hydropower station]. Shuili Shuidian Keji Jinzhan 2007(02):30–33

Tao C, Ma G, Tu Y, Tang M, Guo X (2006a) Zhudian shichang he xuanzhuan beiyong shichang lianhe jingji celüe [Main power market and rotating standby market joint bidding strategy]. Dianli Xitong Jiqi Zidonghua Xuebao 06:10–12

© The Editor(s) (if applicable) and The Author(s) 2025
Y. Tu, *Management of Hydropower Enterprises*, Water Resources Development and Management, https://doi.org/10.1007/978-981-97-5584-4

Tao C, Ma G, Tu Y, Wu S, Liu J, Chen J (2006b) Jiyu hunhe yichuan suanfa de shuidianzhan jingji yunxing [Economic hydropower station operation based on the hybrid genetic algorithm]. Dianli Fadian 2006(05):74–76

Tao C, Ma G (2006) Research on AGC overhead in the electricity market environment. Huadong Dianli 2006(04):5–620

Tao C, Yang Z, He Y, Ma G, Lu L (2012) Daduhe Pubugou yixia tiji shuiku shuisha lianhe diaodu yanjiu [Study of water and sediment cooperative dispatch in cascade reservoirs below Dadu River Pubugou]. Shuili Fadian 38(10):73–75+80

Tu Y et al (2019) Zhihui qiye gailun [Introduction to the intelligent enterprise]. Kexue Chubanshe

Tu Y et al (2018) Zhihui qiye kuangjia yu shijian [Intelligent enterprise framework and practice]. Jingji Ribao Chubanshe

Tu Y (2017) Jianshe zhihui qiye tuidong guanli chuangxin [Intelligent enterprise development pushing management innovation]. Sichuan Shuili Fadian 36(01):148–151

Tu Y (2018a) Jiyu zizhu chuangxin de zhihui qiye jianshe [Development of the intelligent enterprise based on autonomous innovation]. Qiye Guanli 2018(05):21–22

Tu Y (2018b) Shuju qudong zhihui qiye [The data-driven intelligent enterprise]. Qiye Guanli 2018(02):100–103

Tu Y (2016) Shuidian qiye ruhe jianshe zhihui qiye [How hydropower enterprises can build smart enterprises]. Nengyuan 2016(08):96–97

Wang J, Tao C, Ma G, Yang Z, Huang W (2013) Tiji shuidianzhan changjian AGC xitong yanjiu [Research on cascade hydropower station inter-plant AGC system]. Zhongguo Nongcun Shuili Shuidian (09):134–137+141

Wei Q, Chen S, Huang W, Ma G, Tao C (2021) Liyong suiji senlin huigui de xianhuo shichang chuqing jiage yusuan fangfa [Spot market clearing price prediction method using random forest regression]. Zhongguo Dianji Gongcheng Xuebao 41(04):1360–1367+1542

Wen H, Li J (2014) Jiniu shuidianzhan bati bianxing jiance fangan youhua sheji [Optimal design of dam deformation monitoring scheme for Jiniu hydropower station]. Dongbei Shuili Shuidian 32(11):4–6+71

Zhang S, Chen S, Ma G, Huang W, Tao C (2020) Jiyu DPBIL-SVM de dianli xianhuo shichang chuqingjia yuce yanjiu [Research on spot market clearing price prediction based on the DPBIL-SVM hybrid model]. Shuidian Nengyuan Kexue 38(04):197–200

Zhang S, Jiang D (2021) Gaojingdu dibiao sanwei weiyi zidonghua jiance jishu yanjiu yu yingyong [Research and application of high-precision three-dimensional land surface displacement automatic monitoring technology]. Changjiang Kexueyuan Yuanbao 38(01):66–71

Staceynyx. Mount Emei - Sunrise above the clouds. CC BY-SA 3.0

User: Tianjin 24. June 07 350D 127. CC BY 2.0